ADVANCED
COMMON CORE
MATH
EXPLORATIONS

ADVANCED

COMMON CORE
MATH
EXPLORATIONS

Factors &
Multiples

JERRY BURKHART

Routledge
Taylor & Francis Group

NEW YORK AND LONDON

First published in 2014 by Prufrock Press Inc.

Published 2021 by Routledge
605 Third Avenue, New York, NY 10017
2 Park Square, Milton Park, Abingdon, Oxon OX14 4RN

Routledge is an imprint of the Taylor & Francis Group, an informa business

Copyright © 2014 by Taylor & Francis Group

Cover design by Raquel Trevino and layout design by Allegra Denbo

Notice:
Product or corporate names may be trademarks or registered trademarks, and are used only for identification and explanation without intent to infringe.

ISBN 13: 978-1-0321-4293-7 (hbk)
ISBN 13: 978-1-6182-1261-0 (pbk)

DOI: 10.4324/9781003232742

Table of Contents

A Note to Students

Welcome, math explorers! You are about to embark on an adventure in learning. As you navigate the mathematical terrain in these activities, you will discover that "doing the math" means much more than calculating quickly and accurately. It means using your creativity and insight to question, investigate, describe, analyze, predict, and prove. It means venturing into unfamiliar territory, taking risks, and finding a way forward even when you're not sure which direction to go. And it means discovering things that will expand your mathematical imagination in entirely new directions.

Of course, the job of an explorer involves hard work. There may be times when it will take a real effort on your part to keep pushing forward. You may spend days or more pondering a single question or problem. Sometimes, you might even get completely lost. The process can be demanding—but it can also be very rewarding. There's nothing quite like the experience of making a breakthrough after a long stretch of hard work and seeing a world of new ideas and understandings open up before your eyes!

These explorations are challenging, so you might want to team up with a partner or two on your travels—to discuss plans and strategies and to share the rewards of your hard work. Even if you don't reach your final destination every time, I believe you will find that the journey was worth taking. So gear up for some adventure and hard work . . . and start exploring!

Introduction

This introduction contains general information about the structure of the books and the implementation of the activities in the Advanced Common Core Math Explorations series. For additional information and support, please see the free e-booklet *Advanced Common Core Math Explorations: A Teacher's Guide* that accompanies this series at http://www.routledge.com/Assets/ClientPages/AdvancedCCMath.aspx.

AUDIENCE

Advanced Common Core Math Explorations: Factors and Multiples is designed to support students, teachers, and other learners as they work to deepen their understanding of middle school math concepts. The activities have been written primarily with upper elementary and middle school students and teachers in mind. However, older students or those who have already studied more advanced content can also enjoy and benefit from them. The explorations can be used in classrooms, as professional development activities for mathematics teachers, in college math content and methods courses, and by anyone who would like to extend his or her understanding of middle school mathematics concepts by solving challenging problems.

These explorations are designed to stretch students beyond their initial level of comfort. They are built around the belief that most of us underestimate the mathematics we are capable of learning. Although the activities are challenging, they are also meant to be accessible. Although they are targeted to the special needs of gifted and talented students, I hope that teachers will make them available to any student who would like to pursue the challenge. Most students are capable of making progress and learning something meaningful, even if they work just on the first question or two of an activity.

PURPOSE

The investigations in this series were developed through years of work with talented middle school math students. They are designed to:
  » engage students in the excitement of mathematical discovery;
  » deepen students' understanding of a wide range of middle school math concepts;

DOI: 10.4324/9781003232742-1

» encourage the use of multiple strategies for solving problems;
» help students become flexible, creative, yet disciplined mathematical thinkers;
» improve mathematical communication skills;
» highlight connections between diverse mathematical concepts;
» develop perseverance, patience, and stamina in solving mathematical problems;
» provide levels of depth and challenge to meet a variety of needs and interests;
» enable students to work both collaboratively and independently; and
» offer opportunities for further exploration.

STRUCTURE OF THE BOOKS

Each book in the Advanced Common Core Math Explorations series contains ready-to-use explorations focused on one mathematical content area. The content and structure are built around the Common Core State Standards for Mathematics (National Governors Association Center for Best Practices & Council of Chief State School Officers [NGA & CCSSO], 2010), both the Mathematical Content standards and the Standards for Mathematical Practice. Because the emphasis is on challenge and depth, there is a stronger focus on concepts than on procedural skills. However, most activities provide plenty of opportunities to practice computational skills as well.

Each exploration is matched with one or more Common Core benchmarks or clusters, which come with grade-level designations (see p. 8). This grade-level information should serve as a rough guide. When selecting activities, use your own knowledge of your students' backgrounds and abilities. Information about the prior knowledge needed for each exploration is also included as a guide.

FEATURES OF THE EXPLORATIONS

Each activity includes three stages. Stage 1 (and sometimes part of Stage 2) may be challenging enough to meet the needs of many students. The second and third stages are usually appropriate for older students, or for those who finish early, need more challenge, or are highly motivated and curious to learn more. They may also be useful for teachers or other adults who have more mathematical experience and want to extend their own knowledge further. I have separated the explorations into stages in order to provide a tool for setting goals, to help measure and celebrate students' progress, and to create additional options for those who need them.

Each exploration also contains features carefully designed to support teachers in the implementation process: an introduction, the student handout, a teacher's guide with questions and notes to guide conversation and detailed solutions, and suggestions for a closing discussion.

IMPLEMENTING THE EXPLORATIONS

Implementing each exploration involves five steps on the part of the teacher: prepare, introduce, follow up, summarize, and assess.

Prepare

The best way to prepare to teach an activity is to try it yourself. Although this involves an initial time investment on your part, it pays great dividends later. Doing the activity, ideally with a partner or two, will help you become familiar with the mathematics, anticipate potential trouble spots for students, and plan ways to prepare students for success. After you have used the activity once or twice with students, very little preparation will be needed.

Introduce

The Introduction section at the beginning of each exploration provides support to help you get your students started: materials and prior knowledge needed, learning goals, motivational background, and suggestions for launching the activity.

Read the Motivation and Purpose selection to students, and then follow the suggestions for leading a discussion to help them understand the problem. Often, one of the suggestions involves looking through the entire activity with them (or as much of it as they will be doing) to help them see the big picture before they begin. Let students know what kind of a time frame you have in mind for the exploration. An activity may take anywhere from a few days to 2 or 3 weeks depending on how challenging it is for students, how much of it they will complete, and how much class time will be devoted to it.

The explorations are designed to allow students to spend much of their time working without direct assistance. However, it's usually best if you stay with them for a few minutes just after introducing an activity to ensure that they get started successfully. This way, you can catch potential trouble spots early and prevent unnecessary discouragement.

This is also a good time to remind students about the importance of giving clear, thorough written explanations of their thinking. Specific motivation techniques and suggestions for developing mathematical communication skills are included in the *Advanced Common Core Math Explorations: A Teacher's Guide* e-booklet.

Follow Up

The level of challenge in these explorations makes it impractical for most students to complete them entirely on their own as seatwork or homework. Students' most meaningful (and enjoyable!) experiences are often the opportunities you give them to have mathematical conversations with you and with each other while the activity is in progress. If you are implementing an activity with a small group of students in a mainstream classroom, it may be sufficient to plan to meet with them a couple of times per week, for 15 or 20 minutes each time. If circumstances allow more time than this, then the conversations and learning can be still better.

The Teacher's Guide for each exploration reprints each problem and contains two main elements: (a) Questions and Conversations and (b) Solutions. The Questions and Conversations feature is designed to help you facilitate these conversations with and among students. For the most part, it lists questions that students may ask or that you may pose to them. Ideas for responding to the questions are included. It isn't necessary to ask or answer all of the questions. Instead, let students' ideas and your experience and professional judgment determine the flow of the conversation. The Solution section offers ideas for follow-up discussions with students as they work. Although the answers in the Questions and Conversations sections are often intentionally incomplete or suggestive of ideas to consider, you'll find detailed answers, often with samples of multiple approaches that students pursue in the Solution section.

Summarize

After students have finished an exploration, plan a brief discussion (20 minutes is usually enough) to give them a chance to share and critique one another's ideas and strategies. This is also a good time to answer any remaining questions they have. The Wrap Up section at the end of each exploration offers ideas for this discussion, along with suggestions for further exploration.

Assess

One of the most valuable things you can do for your students is to comment on their work. You don't have to write a lot, but your comments should show that you have read and thought about what they have written. Whether you give praise or offer suggestions for growth, make your comments specific and sincere. Ideally, some of your comments will relate to the detail of the mathematical content. Some specific suggestions are included in the free e-booklet accompanying this series.

If you would like to give students a numerical score, consider using a rubric such as the one in *Extending the Challenge in Mathematics: Developing Mathematical Promise in K–8 Students* (Sheffield, 2003). Whatever system you use, the emphasis should be on process goals such as problem solving, reasoning, communication, and making connections—not just correct answers. You may also build in general criteria such as effort, perseverance, correct spelling and grammar, organization, legibility, etc. However, remember that the central goal is to develop students' mathematical capacity. Any scoring system should reflect this.

GETTING STARTED

Here are some tips for getting started. First, a few "DON'Ts" to help you avoid some common pitfalls:

» *Don't feel that you have to finish the activities.* Students will learn more from thinking deeply about one or two questions than from rushing to finish an activity. Each exploration is designed to contain problems that will challenge virtually any student. Most students will not be able to answer every question.

» *Don't feel that you have to explain everything to students.* Your most important job is to help them learn to develop and test their own ideas. They will learn more if they do most of the thinking.

» *Don't be afraid to allow students to struggle.* Talented students need to know that meaningful learning takes time and hard work. Many of them need to experience some frustration—and learn to manage it.

» *Don't feel that you have to know all of the answers.* In order to challenge our students mathematically, we have to do the same for ourselves. You'll never know all of the answers, but if you're like me, you'll learn more about the math every time you teach an exploration! Do what you can during the time you've allotted to prepare, and then allow yourself to learn from the mathematical conversations—right along with your students.

And now some important "DOs":

» *Take your time.* Allow the students plenty of time to think about the problems. Take the time to explore the ideas in depth rather than rushing to get to the next question.

» *Play with the mathematics!* To many people's surprise, math is very much about creative play. Of course, there are learning goals, and it takes effort, but also be sure to enjoy playing with the patterns, numbers, shapes, and ideas!

» *Listen closely to students' ideas and expect them to listen closely to each other.* Meaningful mathematical conversation may be the single most important key to students' learning. It is also your key to assessing their learning.

» *Help students feel comfortable taking risks.* When you place less emphasis on the answers and show more interest in the quality of students' engagement, ideas, creativity, and questions, they will feel freer to make mistakes and grow from them.

» *Believe that the students—and you—can do it!* Middle school students have great success with these activities, but it may take some time to adjust to the level of challenge.

» *Use the explorations flexibly.* You don't always have to use them exactly "as is." Feel free to insert, delete, or modify questions to meet your students' needs. Adjust due dates or completion goals as necessary based on your observations of students.

Teachers who use the activities in a mainstream classroom often find it help-ful to make a solid but realistic commitment at the beginning of the school year to implement the explorations. Put together a general plan for selecting students, forming groups, creating time for students to work (including time for you to meet with them), assessing the activities, and communicating with parents. Stick with your basic plan, making adjustments as needed as the school year progresses.

THE E-BOOKLET

The Advanced Common Core Math Explorations series comes with a free e-booklet (see http://www.routledge.com/Assets/ClientPages/AdvancedCCMath. aspx) that contains detailed suggestions and tools for bringing the activities to life in your classroom. It addresses topics such as motivation, questioning techniques, mathematical communication, assessment, parent communication, implementing the explorations in different settings, and identification.

Connections to the Common Core State Standards

COMMON CORE STATE STANDARDS FOR MATHEMATICAL CONTENT

Table 1 outlines connections between the activities in *Advanced Common Core Math Explorations: Factors and Multiples* and the Common Core State Standards for Mathematics (NGA & CCSSO, 2010). The Standard column lists the CCSS Mathematical Content standards that apply to the activity. Connections shows other standards that are also addressed in the exploration. Extending the Core Learning describes how the activity extends student learning relative to the listed standard(s).

COMMON CORE STATE STANDARDS FOR MATHEMATICAL PRACTICE

The Common Core State Standards for Mathematical Practice are central to purpose and structure of the activities in *Advanced Common Core Math Explorations: Factors and Multiples*. The list below outlines the ways in which the activities are built around these standards, providing a few specific examples for purposes of illustration.

1. **Make sense of problems and persevere in solving them.** All of the explorations in the *Advanced Common Core Math Explorations: Factors and Multiples* book engage students in understanding and solving problems. The process begins when you introduce the activity to your students and have a discussion in which everyone works together to clarify the meaning of the question and think about how to begin. Throughout each exploration, students devise problem-solving strategies, and make and test conjectures to guide their decisions and evaluate their progress as they work. They use visual models such as "building blocks" to represent prime factorizations and physical situations such as liquid measurement and paths of pool balls to help them develop a deep understanding of the underlying concepts. To promote perseverance, the activities have a high level of cognitive demand, and there is support for the teacher and student in the form of motivation strategies, a tiered structure for the explorations, and suggestions for facilitating mathematical conversation.

DOI: 10.4324/9781003232742-2

TABLE 1

Alignment With Common Core State Standards for Mathematical Content

Exploration	Standard	Connections	Extending the Core Learning
1. Building Blocks	4.OA.B.4		Recognize and calculate prime factorizations. Analyze patterns between prime factorizations.
2. 1,000 Lockers	4.OA.B.4		Analyze and explain patterns in factors of square numbers.
3. Factoring Large Numbers	4.OA.B.4		Increase fluency with calculating prime factorizations. Understand relationships between factors of a number.
4. Factor Scramble	4.OA.B.4		Understand connections between factors and prime factors.
5. How Many Factors?	4.OA.B.4	6.EE.A.1 6.EE.A.2.A 8.EE.A.1	Use prime factorizations to count the factors of numbers. Develop and apply a formula.
6. Differences and Greatest Common Factors	6.NS.B.4	4.NF.A.1	Understand the relationship between differences and GCFs. Apply it to streamline the process of fraction simplification.
7. A Measurement Dilemma	6.NS.B.4	6.EE.A.1 6.EE.A.2	Analyze and apply connections between GCF (x,y) and the linear combinations $ax - by$
8. Paper Pool	6.NS.B.4	6.RP.A.1 6.RP.A.3.A	Discover, visualize, and analyze connections between GCFs and LCMs.
9. The GCF-LCM Connection	6.NS.B.4		Develop, justify, and apply a formula relating GCFs and LCMs.
10. Mathematical Mystery Code	4.OA.B.4	6.EE.A.1 7.EE.A.2 8.EE.A.1	Use prime factorizations to explore and justify properties of exponents.

2. **Reason abstractly and quantitatively.** The activities in this book provide students with frequent opportunities to understand and investigate connections between mathematical concepts and quantities, including factors, multiples, prime factorizations, greatest common factors, least common multiples, and exponential expressions. For example, in "Differences and Greatest Common Factors," students are called on to investigate concrete examples of the relationship between differences and greatest common factors, abstract general patterns from them, and represent their discoveries in verbal, pictorial, and symbolic form. In other activities, students are also asked to create and manipulate multiple representations of prime factorizations as a means of making sense of a variety of concepts from number theory.

3. **Construct viable arguments and critique the reasoning of others.** These activities often prompt students to use what they have learned in earlier questions or explorations to justify a conclusion or explain why a new fact must be true. For example, in "Factoring Large Numbers," students use earlier observations about prime factorizations of consecutive natural numbers to explain why there must be an infinite number of primes. In "1000 Lockers," students explore inductive and deductive reasoning by gathering data and observing patterns, making conjectures, and then looking at their process from a new perspective to give logical explanations for the causes of the patterns. The "Questions and Conversations" and "Wrap Up" features in each exploration provide ongoing support for the teacher to lead discussions in which students compare and critique strategies and arguments of others.

4. **Model with mathematics.** In the "Measurement Dilemma" activity, students begin by designing models that apply to a measurement situation. They are then asked to describe other phenomena that could be investigated using the same model. As an extension to the exploration, they are encouraged to use the model to explore connections to musical keys and scales. In "Paper Pool," students study paths of pool balls and then explain how to design tables meeting certain specifications.

5. **Use appropriate tools strategically.** Throughout these explorations, students develop and use a set of tools for finding prime factorizations: mental math, paper and pencil, calculator, and a visual "building blocks" grid. With guidance, they practice choosing the appropriate tool based on the size of numbers involved, their own current level of knowledge and experience, and the complexity of the task at hand.

6. **Attend to precision.** Students are consistently expected to give clear and complete explanations of strategies and procedures in these activities. They correctly use terms such as *factor*, *product*, *multiple*, *prime factorization*, *exponent*, *base*, *greatest common factor*, and *least common multiple* when making and justifying conclusions. (*Note.* You will see that I have italicized terms in the lessons when they appear to help you draw your students' atten-

tion to this vocabulary.) Teachers are provided with support for leading discussions that develop students' communication skills. A section in the e-booklet accompanying the series is devoted to helping students understand why it is important to communicate clearly and precisely and how to do so effectively.

7. **Look for and make use of structure.** Pattern and structure are central components of the explorations in *Advanced Common Core Math Explorations: Factors and Multiples.* By paying close attention to structure in the relationships within and between prime factorizations, students (a) develop efficient strategies for calculating them, (b) discover formulas for counting factors, (c) reason about relationships between the factors of a number, and (d) discover connections between greatest common factors and least common multiples. They regularly stop to evaluate what they are doing as they work, shifting perspectives to gain new understandings. For example, after using patterns in the "Building Blocks" visualization to find factors of a number, they reimagine this representation in terms of exponents. By observing how the patterns express themselves in this different form, they deepen their understanding of both exponents and factors.

8. **Look for and express regularity in repeated reasoning.** In *Advanced Common Core Math Explorations: Factors and Multiples*, students are constantly engaged in calculations and processes that display regularity. They use this predictability to find more efficient procedures, develop equations, and probe connections between concepts. For example, in "How Many Factors?", students design a way to organize the process of finding factors. By noticing similarities in this process for different numbers, they develop a formula that quickly counts the total number of factors. In "Factoring Large Numbers," students gradually discover that composite numbers and numbers larger than the square root of a number need not be tested when searching for factors. They use this observation to streamline the process of finding prime factorizations.

Exploration 1

Building Blocks

Materials

- » Colored pencils

Prior Knowledge

- » Identify and find factors and multiples of one- and two-digit numbers.
- » Understand the concept of a *prime number* (optional, but recommended).

Learning Goals

- » Understand prime numbers as building blocks of the natural (counting) numbers.
- » Analyze connections between prime factorizations of different numbers.
- » Analyze and extend complex patterns.
- » Persist in solving challenging problems.

> **Teacher's Note.** This activity may be used to teach the concept of prime factorization before introducing procedures such as using factor trees. If students already know about prime factorizations, this activity will give them a chance to explore the concept in more depth.

Launching the Exploration

Motivation and purpose. To students: This exploration might not really look like mathematics to you, but actually, it's all about exploring patterns—and that's math! You will eventually discover that some of the math here is familiar, but it's hidden, and part of your job is to find it! Be prepared to see this colored-block grid again in future explorations.

Understanding the problem. Read the directions for Stage 1 to make sure students understand them, but be careful not to give anything away! Don't tell them that it is about factors, multiples, or prime numbers. They will discover this for themselves! Encourage them to record their ideas on separate "thinking" paper until they are confident in the results before transferring their work to the final copy. Offer assistance as needed for a least a few minutes as they begin the exploration to make sure they understand the main idea.

DOI: 10.4324/9781003232742-3

STUDENT HANDOUT

Stage 1

1. Analyze patterns in the colored blocks for the top 50 squares of this grid. Extend the patterns to fill in the remaining 50 squares.

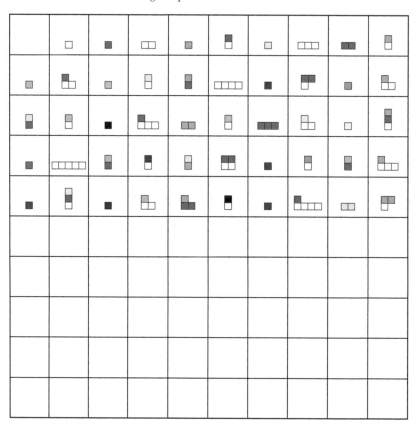

Note. The goal is to figure out how many blocks of each color are in each square, not how they are arranged. (There is a pattern in the arrangement, too, but its purpose is just to make the grid easier to read.)

Stage 2

2. Draw the correct block pattern into the two empty squares and fill in the correct colors for all blocks in the grid. Include any additional information that you think is important.

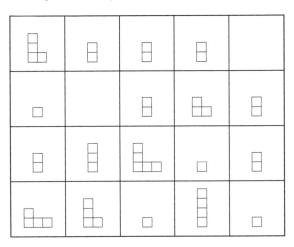

Note. All patterns from the original grid continue to apply. But now, the arrangement of blocks within each square is important!

Stage 3

3. Look at the grid in Stage 1 and focus on the squares that have just one block. Can you find a method that will always predict when the next one will come?

TEACHER'S GUIDE

STAGE 1

Problem #1

1. Analyze patterns in the colored blocks for the top 50 squares of this grid. Extend the patterns to fill in the remaining 50 squares.

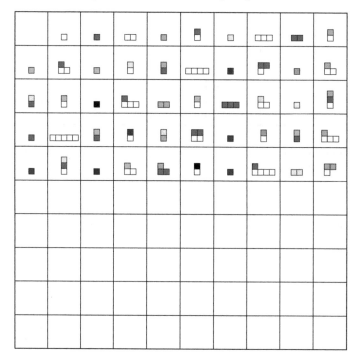

Note. The goal is to figure out how many blocks of each color are in each square, not how they are arranged. (There is a pattern in the arrangement, too, but its purpose is just to make the grid easier to read.)

Questions and Conversations for #1

This section contains ideas for conversations, including questions that students may ask or that you may pose to them. Give them plenty of time to explore on their own before going too deeply into the questions and ideas here. Be sure that students are doing most of the thinking and talking!

»   *Is it important to keep the blank square in the upper left corner?* Yes. The pattern will be easier to understand if you keep it there.

»   *Are there patterns in rows, columns, diagonals, etc.?* Yes, there are many vertical and diagonal patterns. For example, white blocks appear in alternating columns. What pattern do you see for red blocks?

> **Teacher's Notes.** Most students can easily spend a few days on Stage 1. If they finish while others are still working, they can continue to look for more patterns or move on to Stage 2. There are many "small" patterns that can help students fill in parts of the remaining squares. There is also one "general" pattern that ties all of the smaller patterns together.

»   *Does every color show some pattern?* Yes, every color follows some distinctive pattern. Try to think about what might cause these patterns.

»   *Will it help to count the total number of blocks in each row or column?* You might make interesting observations by counting these, but it will probably not help you complete the grid.

»   *Will it help to look for a pattern in the number of white blocks in the squares?* It might, but it's a complicated pattern. Reading from left to right, top to bottom, it looks like this: 1, 2, 1, 3, 1, 2, 1, 4, 1, 2, 1, 3, 1, 2, 1, 5, 1, 2, 1, 3, 1, 2, 1, 4, . . . Every natural number will eventually appear in this list! You might find the pattern hard to unravel without more information.

»   *Can it help to focus on patterns in squares in which all blocks are the same color?* Yes. This will probably help the most for squares with white blocks because you can't see many squares of this type for the other colors.

> **Teacher's Note.** Students often enjoy working on the 1, 2, 1, 3, 1, 2, 1, 4, . . . pattern, but if they have explored it for a while without success, you might suggest that they pursue other avenues. Encourage them to come back to it after they have solved the entire puzzle. By the way, other colors show similar types of complex patterns!

»   *Should I look at the squares in a particular order?* Yes, try reading the grid from left to right, top to bottom, just as you would read a book.

»   *Will two different squares ever have exactly the same collection of colored blocks?* No.

» *What always happens the first time a new color appears?* Every color appears by itself the first time.

» *What always happens the second time a new color appears?* Every color appears with a white block the second time. Can you see how this pattern continues?

» *Does the main pattern have anything to do with numbers?* Yes. Try numbering the squares in order from left to right, top to bottom.

» *Should the first square be numbered "0" since it contains no blocks?* This is an important question. Experiment before deciding. (It turns out that the blank square at the beginning should be numbered 1, not 0.)

» *Look at the squares numbered 2, 3, and 6 or 3, 4, and 12. What happens?* Look at how the block diagrams combine. Under what circumstances does this happen in general?

» *In which square will the next white block go? What about the next red block?* Students might notice that white blocks occur every 2 squares, red blocks come every 3 squares, orange ones appear every 5 squares, etc. (This is a key pattern.)

» *Once you know that a square contains a certain color, how can you figure how many blocks of that color it has?* Look at the squares that have at least two white blocks. What about three white blocks? Ask the same types of questions for other colors.

» *Is there anything special about the squares with just one block?* Yes! Try writing down the numbers of these squares. (These squares are all labeled as prime numbers!)

Teacher's Note. If students discover that some squares never seem to get filled with any blocks, it might be a good time to mention that new colors will sometimes be needed. When this happens, students may choose the color.

Solution for #1

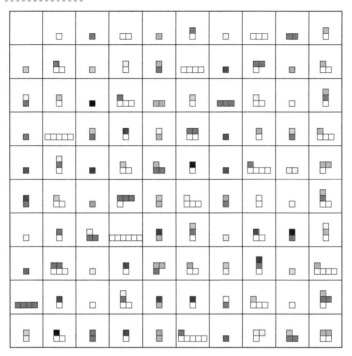

The general pattern is that the completed grid is a picture of the prime factorizations of the natural (counting) numbers from 2 through 100! (The number 1 is represented as no blocks.) Each color represents a prime number: 2 is white, 3 is red, 5 is orange, 7 is yellow, etc. Attaching the blocks represents multiplying the corresponding prime numbers.

STAGE 2

Problem #2

2. Draw the correct block pattern into the two empty squares and fill in the correct colors for all blocks in the grid. Include any additional information that you think is important.

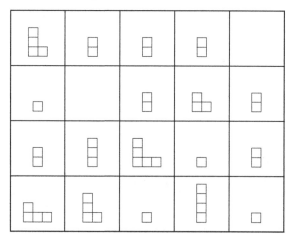

Note. All patterns from the original grid continue to apply. But now, the arrangement of blocks within each square is important!

Questions and Conversations for #2

» *How are the blocks arranged in each square?* Each row of blocks contains only one prime number value. The smallest prime number is in the bottom row. The values of the prime numbers increase as you ascend.

» *Does each square still represent a number?* Yes. The 20 squares represent 20 consecutive natural numbers. (The *natural numbers* are the counting numbers: 1, 2, 3, 4, 5, . . . This may be a good time to introduce this vocabulary to students.)

» *Does the first square have a final digit of 1?* Not necessarily.

» *Can I just choose a number for one of the boxes and see if it works?* That might be a good place to start. Pay attention to what happens when you do this. It might eventually help you find more efficient methods.

» *Will the patterns in the columns and diagonals be the same as before?* Think about the fact that there are now five columns instead of 10. How does this affect these patterns?

» *Where are the prime numbers in the grid? How can they help?* As before, these are the squares that have just one block. How can you combine this knowledge with an understanding of the spacing between different colors?

Solution for #2

444 $2^2 \cdot 3 \cdot 37$	445 $5 \cdot 89$	446 $2 \cdot 223$	447 $3 \cdot 149$	448 $2^6 \cdot 7$
449 449	450 $2 \cdot 3^2 \cdot 5^2$	451 $11 \cdot 41$	452 $2^2 \cdot 113$	453 $3 \cdot 151$
454 $2 \cdot 227$	455 $5 \cdot 7 \cdot 13$	456 $2^3 \cdot 3 \cdot 19$	457 457	458 $2 \cdot 229$
459 $3^3 \cdot 17$	460 $2^2 \cdot 5 \cdot 23$	461 461	462 $2 \cdot 3 \cdot 7 \cdot 11$	463 463

Squares shown as light grey in this key may be any color that hasn't been used yet, but students should make each one a different color because each represents a different prime number. The color they use for 89 in the block diagram for 445 should be the same as the one they chose in the original grid.

Teacher's Note. Because the question did not specify the additional information that students should include, they may not have shown everything in this grid, especially the expressions for the prime factorizations. These are included here mainly for your information.

STAGE 3

Problem #3

3. Look at the grid in Stage 1 and focus on the squares that have just one block. Can you find a method that will always predict when the next one will come?

Questions and Conversations for #3

» *Why are there so many patterns that work for quite a while and then fall apart?* This can be one of the great frustrations and challenges of working with prime numbers!

Teacher's Note. Encourage students to explore beyond 100! Pay attention to the separation between the prime numbers. How many prime numbers are there in certain intervals? What seems to happen to this density of primes as the numbers increase? Why does this happen?

Solution for #3

This is an unsolved problem in mathematics! Mathematicians have been working on it for centuries. However, students may make some interesting observations. For example, prime numbers tend to get more spread out as numbers get larger. No matter how large a number you choose, it is always possible to find two consecutive prime numbers that are at least that far apart.

WRAP UP

Share Strategies

Ask students to share patterns, strategies, and observations from their work on Stage 1. Make connections between the different patterns. If some students have worked on Stage 2, have them share strategies and discoveries for this as well.

Students who have worked on Stage 3 may be surprised that you have asked them to solve a problem to which no one knows the answer! This is a great opportunity to get them accustomed to the idea that there are still unsolved problems in mathematics (and that correct answers are not the only goal in math class).

Summarize

If no one has discovered the general pattern, share it with students, introducing the term *prime factorization* if necessary. Be sure to acknowledge any progress they've made. Discuss how the general pattern causes the other patterns.

If it hasn't come up yet, introduce the term *natural number*. The natural (counting) numbers are the numbers 1, 2, 3, 4, 5, . . .—all of the numbers that you count with, beginning with 1. They don't include the number 0. The grid shows the prime factorizations of the natural numbers from 2–100. The number 1 technically doesn't have a prime factorization because it is not defined to be a prime number. (Some people might say that it has an "empty" prime factorization. This is reflected in the fact that its square contains no blocks!)

Further Exploration

Ask students to think of new questions to ask. Here are some possibilities:
» How could you find a colored block diagram for a number much larger than 100 without coloring in diagrams for all of the smaller natural numbers in between?
» Is it possible that there is a natural number other than 1 that does not have a colored block diagram (prime factorization)? Is it possible to find a number that has two different colored block diagrams? How can the block diagrams help you find some or all of the factors of a number?
» Try creating your own puzzle like the one in Stage 2. Will there always be a solution? How many of the squares do you need to fill in to ensure that there is only one solution?

For more ideas on using the colored blocks model to represent prime numbers, see the article *Building Numbers From Primes* (Burkhart, 2009).

Exploration 2

1000 Lockers

INTRODUCTION

Prior Knowledge

- » Identify and find factors and multiples of one- and two-digit numbers.
- » Understand the concept of a *square number*.

Learning Goals

- » Understand the relationship between factors and multiples.
- » Gather and organize a complex set of data.
- » Recognize and extend patterns.
- » Make and prove a conjecture about factors of square numbers.
- » Think flexibly and use multiple strategies to solve a problem.
- » Use inductive and deductive reasoning.
- » Communicate complex mathematical ideas clearly.
- » Persist in solving challenging problems.

Launching the Exploration

Motivation and purpose. To students: The 1000 Lockers problem is a popular mathematics puzzle that has entertained and challenged people for years. It will develop your skills in the areas of problem solving, organizing complex information, managing detail, discovering patterns, making predictions, and using logic to unearth hidden causes of patterns.

Understanding the problem. Read through the problem together. Pose questions to help students visualize the situation:

- » *What will happen after the first person walks all the way through?* All of the lockers will be open.
- » *The second person?* All even-numbered lockers will be closed. All odd-numbered lockers will be open.
- » *The third person?* Now it starts to get complicated!
- » *How can you keep track of what is happening?* Students might suggest a way of organizing their work, such as using a table.

DOI: 10.4324/9781003232742-4

STUDENT HANDOUT

Imagine a row of 1000 lockers, numbered in order from 1 to 1000. All of the lockers are closed. A line of 1000 students approaches the lockers. The first student opens every locker that is a multiple of 1. The second student *changes* every locker that is a multiple of 2. (If it's open, she closes it, and if it's closed, she opens it.) The third person in line changes every locker that is a multiple of 3. This process continues until all 1000 people have gone by. Which lockers will be open at the end?

Stage 1

1. Design a system to keep track of what happens as the people walk through. Use it to solve the problem.

Stage 2

2. Solve the problem a different way. Check that your solutions agree.

3. Use one of your solutions to explain what causes the patterns that you found.

Stage 3

4. What happens if only the even-numbered people walk by? The odd-numbered people?

TEACHER'S GUIDE

Imagine a row of 1000 lockers, numbered in order from 1 to 1000. All of the lockers are closed. A line of 1000 students approaches the lockers. The first student opens every locker that is a multiple of 1. The second student *changes* every locker that is a multiple of 2. (If it's open, she closes it, and if it's closed, she opens it.) The third person in line changes every locker that is a multiple of 3. This process continues until all 1000 people have gone by. Which lockers will be open at the end?

STAGE 1

Problem #1

1. Design a system to keep track of what happens as the people walk through. Use it to solve the problem.

Questions and Conversations for #1

This section contains ideas for conversations, mainly in the form of questions that students may ask or that you may pose to them. Be sure to allow students to do most of the thinking and talking!

» *What makes the problem seem difficult?* Most people would say that it's the large number of lockers that makes it hard.

» *Is there a way to get around this difficulty?* You could start with a smaller number of lockers and use this to understand the problem better. More lockers can always be included later.

» *Would it help to act the problem out?* Maybe! Think: Would this make it easier to solve the problem? Would it make it more interesting? Is it practical? Is there a way to act it out without actually using lockers?

» *How many lockers should you start with?* What is practical? How many might be enough to help you understand the problem? (Students typically settle on 10 or 20 lockers. It's fairly quick to do 10 and that might be enough to suggest a pattern. On the other hand, using more lockers might increase your confidence in any predictions you make.)

» *Does it matter how many students walk through?* Think about how the number of students should relate to the number of lockers. (Students usually discover that you never need more people than lockers.)

» *What is the best way to organize the information?* Many people use a table, but it may depend on how you are thinking about the problem.

» *Why is it important to work carefully?* Errors will lead to further incorrect results later. Also, if you make a mistake, you are unlikely to find a pattern. On the other hand, if you record your work thoroughly and organize it

25

carefully, you should be able to track down and correct errors without too much trouble.

» *Is there more than one way to see the pattern?* Yes! You may focus on the number of lockers between open lockers. Or you may think of what you have to add to an open locker's number to get the number of the next open locker. Some people might notice that the open lockers are labeled with a certain type of number.

Solution for #1

The open lockers will be: 1, 4, 9, 16, 25, 36, 49, 64, 81, 100, 121, 144, 169, 196, 225, 256, 289, 324, 361, 400, 441, 484, 529, 576, 625, 676, 729, 784, 841, 900, and 961. Students might make tables that look something like this:

	Locker									
Student	1	2	3	4	5	6	7	8	9	10
1	**O**	O	O	O	O	O	O	O	O	O
2	O	**C**	O	C	O	C	O	C	O	C
3	O	C	**C**	C	O	O	O	C	C	C
4	O	C	C	**O**	O	O	O	O	C	C
5	O	C	C	O	**C**	O	O	O	C	O
6	O	C	C	O	C	**C**	O	O	C	O
7	O	C	C	O	C	C	**C**	O	C	O
8	O	C	C	O	C	C	C	**C**	C	O
9	O	C	C	O	C	C	C	C	**O**	O
10	O	C	C	O	C	C	C	C	O	**C**

Lockers 1, 4, and 9 are open after 10 students have gone through. (Grey letters show lockers that aren't changing any more.)

Here are some patterns that students might see:

» The open lockers begin at 1 and increase by successive odd numbers.

$$1 + 3 = 4 \qquad 4 + 5 = 9 \qquad 9 + 7 = 16 \qquad 16 + 9 = 25$$

» The spaces between consecutive open lockers contain even numbers of lockers in the following pattern:

2 closed lockers   4 closed lockers   6 closed lockers

» The open lockers are the square numbers.

STAGE 2

Problem #2

2. Solve the problem a different way. Check that your solutions agree.

Questions and Conversations for #2

» *What happens with the prime-numbered lockers? Why?* They will end up being closed because they will be changed twice—opened once by the first person and then closed by the person whose number is the same as the locker number.

» *Can you apply a similar type of thinking to other lockers?* Yes. Try focusing on one locker at a time.

Solution for #2

Look at the factors of the locker numbers.

Locker Number	Factors
1	1
2	1, 2
3	1, 3
4	1, 2, 4
5	1, 5
6	1, 2, 3, 6
7	1, 7
8	1, 2, 4, 8
9	1, 3, 9
10	1, 2, 5, 10

Locker numbers with an even number of factors will be closed. Those with an odd number of factors will be open. So far at least, this appears to agree with our earlier solution!

27

Problem #3

3. Use one of your solutions to explain what causes the patterns that you found.

Questions and Conversations for #3

» *What do these pictures represent? How do they relate to "1000 Lockers"?* They show that the sums 1, $1+3$, $1+3+5$, and $1+3+5+7$ create square numbers. How can you see this?

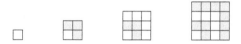

» *How can a picture like this help you predict what happens to one of the lockers?* It tells you something about what will happen to locker number 4. How?

Solution for #3

Because our solutions agree, we suspect that the square numbers must be the numbers that have an odd number of factors. Why? Factors come in pairs. Because of this, we might expect that all numbers will have an even number of factors. However, for square numbers, one of the factors is paired with itself, resulting in an odd number of factors.

We can use pictures to show this. Some students might be familiar with these "factor rainbows." For example, 24 has an even number of factors because each of its factors pairs with another. However, because 36 is a square number, it has an odd number of factors because the number 6 pairs with itself.

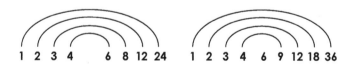

STAGE 3

Problem #4

4. What happens if only the even-numbered people walk by? The odd-numbered people?

Questions and Conversations for #4

» *How can you use what you learned in earlier problems? What will change? What will stay the same?*

Solution for #4

Even-numbered people: The open lockers will be the doubles of the square numbers

$$2, 8, 18, 32, 50, 72, 98, 128, 162, 200, 242, 288, 338, 392,$$
$$450, 512, 578, 648, 722, 800, 882, \text{ and } 968$$

because these are the numbers that have an odd number of even factors.

Odd-numbered people: The open lockers will be all of the square numbers

$$1, 4, 9, 16, 25, 36, 49, 64, 81, \ldots$$

and the doubles of square numbers

$$2, 8, 18, 32, 50, 72, 98, 128, 162, \ldots$$

because these are the numbers that have an odd number of odd factors.

Teacher's Notes. Students may have written the numbers in a single list like 1, 2, 4, 8, 9, 16, 18, 25, 32, etc. This makes it pretty hard to see a pattern! If they are having trouble, ask them if they see any familiar numbers. If they notice that the square numbers are all in the list, suggest that they write down the remaining numbers in a separate list and look for a pattern there.

Students may be curious about what causes these patterns. Exploration 5 will give them some tools that will help them explore this further!

WRAP UP

Share Strategies

Ask students to share and compare their strategies and observations. Make sure that they connect the different ways of looking at the pattern of open lockers.

Summarize

Answer any remaining questions that students have.

Ask students to explain why it is important to understand why the pattern works. This could lead to a discussion of *inductive* and *deductive* reasoning.

Inductive reasoning involves using examples to find patterns and make conjectures. Students were using inductive reasoning when they looked for a pattern in the open lockers and used it to make a prediction.

Once you have made a conjecture, you often test it by investigating more examples. Mathematicians reason this way all of the time, but it always leaves doubt about the validity of your conjecture. To be certain that your conjecture is always true (or not!), you must use deductive reasoning.

Deductive reasoning involves using logic to prove something. Students used deductive reasoning when they explained why the pattern of square numbers occurs. Once they knew the reason, they could be certain that the pattern would continue.

Further Exploration

Ask students to think of ways to continue or extend this exploration. Here are some possibilities:

» In this exploration, you saw a pattern in the sums of odd numbers:

| 1 | 1+3 | 1+3+5 | 1+3+5+7 |

Explore patterns in other sums. For example:

1	1+2	1+2+3	1+2+3+4
2	2+4	2+4+6	2+4+6+8
1	1+4	1+4+9	1+4+9+16

» What would happen if other subsets of the 1000 people walked past the lockers—for example, people who are multiples of some chosen number?
» Ask the question in reverse. Is it possible to find a subset of people to "send through" that will result in a chosen set of lockers being open?

Exploration **3**

Factoring Large Numbers

INTRODUCTION

Materials

> » Calculator
> » Completed "Building Blocks Grid" (see http://www.routledge.com/Assets/ClientPages/AdvancedCCMath.aspx)

Prior Knowledge

> » Complete Stage 1 of Exploration 1: Building Blocks.
> » Use division to find factors of multidigit numbers.
> » Understand the meaning of *square root*.

Learning Goals

> » Recognize the special challenges involved in factoring large numbers.
> » Develop and apply efficient strategies to find prime factorizations of large numbers.
> » Become more familiar with prime factorizations of numbers between 1 and 100.
> » Analyze relationships between a number's factors. (For example: If 4 is a factor of a number, so is 2. If a factor of a number is less than its square root, then there is a corresponding factor that is greater than its square root.)
> » Communicate complex mathematical ideas clearly.
> » Persist in solving challenging problems.

Launching the Exploration

Motivation and purpose. To students: Very large numbers can be surprisingly hard to factor. In fact, this very difficulty is used to protect people's private information on the Internet. In this exploration, you will encounter a few of the challenges involved in factoring large numbers and you will create your own strategies for overcoming some of them. The numbers you factor won't be nearly as large as the ones used for Internet security, but you will learn a lot about how prime numbers and prime factorizations work!

DOI: 10.4324/9781003232742-5

Understanding the problem. Remind students that a *natural number* is a counting number: 1, 2, 3, 4, etc. Read the Stage 1 questions together and discuss the meaning of the word *efficient*.

Encourage students to use a calculator to test numbers as possible factors. (If they know some divisibility tests, they could use these as well.) Be prepared to assist them for at least a few minutes as they begin work to ensure that they are able get started successfully.

STUDENT HANDOUT

Stage 1

1. Find the prime factorizations of these large numbers: 209, 2275, 1536, and 941. As you work, look for ways to find the answers more efficiently. Describe your strategies.

2. Prove that 3359 is prime. Use the most efficient method you can.

3. Find a prime number that is greater than 10,000. Explain your strategy. Prove that your number is prime.

Stage 2

4. Why will two consecutive numbers on the "Building Blocks" grid never share a block of the same color?

5. What are the next four numbers in this sequence? Explain your answer.

$$1, 2, 6, 30, 210, 2310, \ldots$$

6. Find the prime factorizations of 2, 6, 30, 210, 2310, and the next number in the sequence. Then find the prime factorizations of the numbers that are one less and one greater than each of these numbers. Organize your results in a table.

7. How does your answer to Problem #6 support your conclusion in Problem #4? Explain.

Stage 3

The purpose of the questions in Stage 3 is to understand how we know that there are infinitely many prime numbers. The Greek mathematician, Euclid, presented a proof of this fact in around 300 B.C.

8. Starting at the number 2, list prime numbers in order for as long as you can without skipping any. Don't use a calculator or the colored-block grid. Stop when it takes you more than about 30 seconds to check that a number is prime.

9. Explain how you could find the smallest number that has all of the prime numbers in your list as factors. Write an expression for the number. (You don't actually have to calculate it. It may be too large for your calculator anyway!)

10. How can you find a number that doesn't have any of the prime numbers in your list as a factor? Again, write an expression for the number. Explain your thinking.

11. Explain why every natural number (except 1) must have a prime factorization.

12. Combine your observations from the previous two questions to explain why there must be infinitely many prime numbers.

TEACHER'S GUIDE

STAGE 1

Problem #1

1. Find the prime factorizations of these large numbers: 209, 2275, 1536, and 941. As you work, look for ways to find the answers more efficiently. Describe your strategies.

Questions and Conversations for #1

This section contains ideas for conversations, mainly in the form of questions that students may ask or that you may pose to them. Be sure to allow students to do most of the thinking and talking!

» *What does it mean to work systematically? What are the advantages of doing this?* Systematic work is planned and predictable. In this problem, it might mean testing small numbers first and gradually trying larger ones. This can ensure that you don't miss factors or retest numbers that you've already tried. It might also make it easier to spot patterns.

» *Do you have to use factor trees?* No. A factor tree is just one tool for organizing the work of finding prime factorizations. You can choose any method that works well for you.

» *Does the process get easier after you've found some of the factors?* Yes. If you divide the original number by any of the factors you've found, you will have a smaller number to work with.

» *Can it help to make lists of numbers that you've tested?* Yes. Keeping records of your work can guide your thinking process. It can help you track down errors and plan the next steps in your investigation. You might also notice patterns in the types of numbers that do and don't work.

» *Are there numbers that you don't have to test? Why or why not?* When you find a number that's not a factor, then none of its multiples will be factors either. For example, this list shows that every multiple of 6 is also a multiple of 3. If a number is not divisible by 3, it is not in this list, so it cannot be divisible by 6 either.

   3, **6**, 9, **12**, 15, **18**, 21, **24**, 27, **30**, 33, **36**, 39, **42**, 45, **48**, . . .

» *What types of numbers do you have to test?* Students may gradually discover (because of the answer to the question above) that they only have to test prime numbers.

» *Do you have to test numbers greater than half the number you are factoring?* No. If you multiply these numbers by 2 (the smallest possible prime factor), the answer will be greater than the original number.

» *Can you stop testing numbers before you reach half the original number?* Yes, usually long before then. Either of the next two questions can help you think about why.

» *Can it help to keep track of the quotients as you test numbers?* Yes. Here is an example. When can you stop testing?

Testing the Number 53

$53 \div 2 = 26.5$

$53 \div 3 \approx 17.7$

$53 \div 5 \approx 10.6$

$53 \div 7 \approx 7.6$

$53 \div 11 \approx 4.8$

(You can stop when the quotient, 4.8, becomes smaller than the divisor, 11. Can you see why?)

» *Can it help to draw "factor rainbows"?* It might. These examples show factor rainbows for 36 and 54. Notice that every factor on the left of a diagram pairs with a factor on the right. This means you can stop testing when you reach the center of the rainbow. Why?

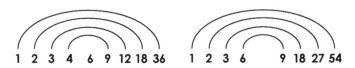

» *What number is at the center of a factor rainbow? Why?* The number at the center is the square root of the number you are testing, because it's the number that you multiply by itself to equal the original number. This means that every factor less than the square root of the number will pair with a factor that is greater than its square root.

» *How accurately do you need to know a number's square root in order to know when you can stop testing the number for factors?* You only need to know what two nearest whole numbers the square root is between.

Solution for #1

$209 = 11 \cdot 19$

$1536 = 2 \cdot 2 \cdot 2 \cdot 2 \cdot 2 \cdot 2 \cdot 2 \cdot 2 \cdot 2 \cdot 3$ or $2^9 \cdot 3$

$2275 = 5 \cdot 5 \cdot 7 \cdot 13$ or $5^2 \cdot 7 \cdot 13$

941 is prime

Sample responses for 1536:

» Use a factor tree.

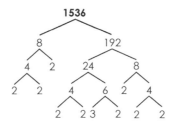

» Keep dividing by 2 until the quotient is no longer even. Because you divide by 2 nine times and the result is 3, there must be nine factors of 2 and one factor of 3 in the prime factorization.

» Divide by 2 until you get an answer that is less than 100.

$$1536 \div 2 = 768$$
$$768 \div 2 = 384$$
$$384 \div 2 = 192$$
$$192 \div 2 = 96$$

Then look at the "Building Blocks" grid to see that the prime factorization of 96 has five factors of 2 and a factor of 3. Combine these with the four factors of 2 that you found at the beginning. This gives a total of nine factors of 2 and one factor of 3.

Sample student response for 941: Only prime numbers need to be tested, because if they don't work, none of their multiples will work. Because $30 \cdot 30 = 900$ and $31 \cdot 31 = 961$, and because 941 is between 900 and 961, the middle of the rainbow diagram must be between 30 and 31. Because 29 is the largest prime number less than or equal to 30, it is the last prime number that you need to test. When you test all of the prime numbers from 2 to 29, none of them are factors of 941. Therefore, it must be a prime number.

Problem #2

2. Prove that 3359 is prime. Use the most efficient method you can.

Questions and Conversations for #2

See Questions and Conversations for #1.

Solution for #2

Sample student response: Use a calculator to find that the square root of 3359 is between 57 and 58. The colored block grid shows that 53 is the largest prime number that is less than or equal to 57. Test all prime numbers from 2 through 53. Because none of them are factors of 3359, it must be prime.

Problem #3

3. Find a prime number that is greater than 10,000. Explain your strategy. Prove that your number is prime.

Questions and Conversations for #3

See Questions and Conversations for #1.

Solution for #3

There are many solutions. 10,007 and 10,009 are the two smallest. Students' answers should show an understanding of the fact that they only have to test prime numbers that are less than or equal to the square root of 10,000. Because $\sqrt{10,000} = 100$, it turns out that the "Building Blocks" grid contains all of the prime numbers they need.

Note: $10,001 = 73 \cdot 137$ and $10,003 = 7 \cdot 1429$.

STAGE 2

Problem #4

4. Why will two consecutive numbers on the "Building Blocks" grid never share a block of the same color?

Questions and Conversations for #4

» *What happens if you focus on one color at a time—for example, the white blocks?* All white blocks are at least two squares apart. What about the other colors?

Solution for #4

Consecutive numbers will never share a common color because white blocks are always at least 2 squares apart, red blocks are always at least 3 squares apart, orange blocks are always at least 5 squares apart, etc. The other colors are even farther apart.

Problem #5

5. What are the next four numbers in this sequence? Explain your answer.

$$1, 2, 6, 30, 210, 2310, \ldots$$

Questions and Conversations for #5

» *Look at the sizes of the numbers. What operation appears to be involved?* The size is increasing more and more dramatically as the list continues. This might suggest the operation of multiplication.

Solution for #5

You multiply by 2, then 3, 5, 7, 11, etc.—the prime numbers. The next four numbers will be

$2310 \cdot 13 = 30,030$ $30,030 \cdot 17 = 510,510$

$510,510 \cdot 19 = 9,699,690$ $9,699,690 \cdot 23 = 223,092,870$

Problem #6

6. Find the prime factorizations of 2, 6, 30, 210, 2310, and the next number in the sequence. Then find the prime factorizations of the numbers that are one less and one greater than each of these numbers. Organize your results in a table.

> **Teacher's Note for #6.** The calculations for the larger numbers will require some patience and persistence. Near the end, it may help students to have a list of the prime numbers between 100 and 200: 101, 103, 107, 109, 113, 127, 131, 137, 139, 149, 151, 157, 163, 167, 173, 179, 181, 191, 193, 197, and 199.

Solution for #6

Let n represent the numbers in the list. Then $n-1$ and $n+1$ are the neighboring numbers.

$n-1$	n	$n+1$
1: no prime factorization	$2 = 2$	3 is prime
5 is prime	$2 \cdot 3 = 6$	7 is prime
29 is prime	$2 \cdot 3 \cdot 5 = 30$	31 is prime
$11 \cdot 19 = 209$	$2 \cdot 3 \cdot 5 \cdot 7 = 210$	211 is prime
2309 is prime	$2 \cdot 3 \cdot 5 \cdot 7 \cdot 11 = 2310$	2311 is prime
30,029 is prime	$2 \cdot 3 \cdot 5 \cdot 7 \cdot 11 \cdot 13 = 30,030$	$59 \cdot 509 = 30,031$

Problem #7

7. How does your answer to Problem #6 support your conclusion in Problem #4? Explain.

Solution for #7

The prime factors of n are never the same as any of the prime factors of $n-1$ or $n+1$. This is expected because neighboring squares in the colored-block grid never share a common color.

STAGE 3

The purpose of the questions in Stage 3 is to understand how we know that there are infinitely many prime numbers. The Greek mathematician, Euclid, presented a proof of this fact in around 300 B.C.

Problem #8

8. Starting at the number 2, list prime numbers in order for as long as you can without skipping any. Don't use a calculator or the colored-block grid. Stop when it takes you more than about 30 seconds to check that a number is prime.

Questions and Conversations for #8

» *How can you know that there are infinitely many prime numbers when it's impossible to keep listing them forever?* You have to use logical thinking to understand why they must have no end.

Solution for #8

Sample student response:

$2, 3, 5, 7, 11, 13, 17, 19, 23, 29, 31, 37, 41, 43, 47, 53, 59, 61$

Of course, different students will stop at different prime numbers.

Problem #9

9. Explain how you could find the smallest number that has all of the prime numbers in your list as factors. Write an expression for the number. (You don't actually have to calculate it. It may be too large for your calculator anyway!)

Questions and Conversations for #9

See Questions and Conversations for #8.

Solution for #9

You can just multiply each of the prime numbers together! In this case the expression would be

$$2 \cdot 3 \cdot 5 \cdot 7 \cdot 11 \cdot 13 \cdot 17 \cdot 19 \cdot 23 \cdot 29 \cdot 31 \cdot 37 \cdot 41 \cdot 43 \cdot 47 \cdot 53 \cdot 59 \cdot 61$$

This is the *smallest* number that has all of the prime factors in the list because it includes no additional prime factors beyond one of each number.

Problem #10

10. How can you find a number that doesn't have any of the prime numbers in your list as a factor? Again, write an expression for the number. Explain your thinking.

Questions and Conversations for #10

See Questions and Conversations for #8.

Solution for #10

Subtract 1 from (or add 1 to) the number in the previous question. The result cannot have any prime factors in common with this number. (See Problem #4.) If you subtract 1, the expression will be

$$2 \cdot 3 \cdot 5 \cdot 7 \cdot 11 \cdot 13 \cdot 17 \cdot 19 \cdot 23 \cdot 29 \cdot 31 \cdot 37 \cdot 41 \cdot 43 \cdot 47 \cdot 53 \cdot 59 \cdot 61 - 1$$

Problem #11

11. Explain why every natural number (except 1) must have a prime factorization.

Questions and Conversations for #11

See Questions and Conversations for #8.

Solution for #11

Imagine filling in squares in the colored-block grid one number at a time in order. Whenever you come to a number that you cannot build from earlier prime factors, it is prime, so you will show it as a single block with a new color. Either way you can build it. In other words, it has a prime factorization.

Problem #12

12. Combine your observations from the previous two questions to explain why there must be infinitely many prime numbers.

Questions and Conversations for #12

» *How can you see that there must be a prime number larger than the largest one in your list (see Problem #8)?* Think about the table you created in Problem #6 and the observations you made in Problems #4 and #10.

Solution for #12

No matter how large a prime number you choose, there must always be one that is larger. We can understand why by looking back.

Let's use N to name the largest number you found in Problem #8. No matter how large N is, you will always be able to use the process in Problems #9 and #10 to find a number that has none of the prime numbers less than or equal to N as a factor.

Because this number can be built (see Problem #11), it must be built from only prime numbers that are larger than N. This means that, no matter how large N is, there must always be a prime number larger than N! In other words, there are infinitely many prime numbers.

WRAP UP

Share Strategies

If students have not already had a chance to do this, allow them to share their methods for finding prime factorizations and to name some advantages and disadvantages of each. Identify and discuss any errors in students' thinking.

One common error is to decompose numbers by addition somewhere in the process. For example, students may think of 209 as $200 + 9$ and combine the prime factorizations of 200 and 9 to get $2 \cdot 2 \cdot 5 \cdot 5 \cdot 3 \cdot 3$. They can catch the error by multiplying it out. This should also help them to see why the addition operation causes the problem.

Summarize

Summarize the main point of the exploration: The only factors you need to test are prime numbers less than or equal to the square root of the number in question. Ask students to explain the reasons for this.

If students have worked on Stages 2 or 3, let them talk about their discoveries.

Further Exploration

Ask students to think of ways to continue or extend this exploration. Here are some possibilities:

» Do some research. How large is the largest known prime number? What other methods have mathematicians discovered for making the factoring process more efficient? Read about the connections between large number factorizations and Internet security.

» Try to find all possible factor trees for a number. Look for patterns and try to predict the number of factor trees in advance. (*Hint*: Consider starting with the "Building Blocks" representation for a number and keep splitting it into pairs until all of the blocks have been completely separated.)

» Increase your fluency with the procedures in the exploration by continuing to practice the methods you have discovered. For example, find prime factorizations of as many natural numbers as you can in order beginning at 101, or of large numbers chosen at random. Pay attention to interesting occurrences such as long stretches of consecutive natural numbers without any primes. Keep thinking of new questions to ask as you work.

Exploration **4**

Factor Scramble

INTRODUCTION

Prior Knowledge

» Use division to find factors of multidigit numbers (with and without calculators).
» Know some techniques for finding prime factorizations.
» Complete Exploration 1: Building Blocks (optional).

Learning Goals

» Build fluency with procedures for finding prime factorizations of large numbers.
» Solve problems and draw conclusions about connections between a number's factors and its prime factorization.
» Communicate complex mathematical ideas clearly.
» Persist in solving challenging problems.

Launching the Exploration

Motivation and purpose. To students: In this exploration, math and word puzzles are blended into one challenging activity! As you work on the puzzles, you will develop a deeper understanding of how a number's factors are connected to its prime factorization.

Understanding the problem. Read through the introductory paragraph. Ask each student to find the value of her or his first name. Then tell students that they will be reversing this process in this exploration.

Find simple words for values such as 65 (me) or 105 (go). Are there any other words that fit these values? How do you know?

Ask students to think about the role of the letter A. They should see that its appearance in a word has no effect on its value. A word may have as many or as few occurrences of A as desired.

Remain available to assist students as they begin the activity. Encourage them to write the prime factorization for each number 2 through 26 next to the corresponding letter of the alphabet. This will simplify their work later in the activity.

45

DOI: 10.4324/9781003232742-6

STUDENT HANDOUT

Stage 1

Every letter is assigned a value as shown below. Words are assigned values by multiplying the values of their letters. For example, the word *zephyr* has a value of $26 \cdot 5 \cdot 16 \cdot 8 \cdot 25 \cdot 18 = 7,488,000$.

A	1	J	10	S	19
B	2	K	11	T	20
C	3	L	12	U	21
D	4	M	13	V	22
E	5	N	14	W	23
F	6	O	15	X	24
G	7	P	16	Y	25
H	8	Q	17	Z	26
I	9	R	18		

Let's find a person's first name with the value 161,000. We'll take the process in steps.

1. Find the prime factorization of 161,000.

2. Why can't the letters C, F, I, L, O, R, U, and X be used? (Try to answer this without testing each letter separately.)

3. Find seven other letters that cannot be used. Explain your thinking.

4. Why must the name contain the letter W?

5. Why must the name contain either G or N, but not both?

6. Find a person's first name with the value 161,000. Explain your thinking.

7. Use prime factorizations to find a word that fits the clue for each value. Prove that your answers are correct.

Value: 83,538	a task at school
Value: 51,129	a good way to exercise
Value: 140,625	a classic toy
Value: 152,460	a person's first name
Value: 43,740	something on a house in winter

Advanced Common Core Math Explorations: Factors & Multiples © Taylor & Francis.
Permission is granted to photocopy or reproduce this page for single classroom use only.

8. Find a number that cannot be the value of any word. Explain your thinking.

9. Many words have values with one or more zeros at the end of the numeral. Why do you think this happens so often?

10. Find the prime factorization of 720. Use it to find at least six words that have this value. Use the prime factorization to show that each word is correct.

11. Suppose each letter is assigned a unique prime number value:

A	2	J	29	S	67
B	3	K	31	T	71
C	5	L	37	U	73
D	7	M	41	V	79
E	11	N	43	W	83
F	13	O	47	X	89
G	17	P	53	Y	97
H	19	Q	59	Z	101
I	23	R	61		

Do you think this makes the puzzle easier or harder? In what other interesting ways does it change the puzzle? Give examples.

Stage 3

12. Show how to use the prime factorization of 4998 to find all of its factors. How can you tell when you have found them all?

Advanced Common Core Math Explorations: Factors & Multiples © Taylor & Francis.

TEACHER'S GUIDE

STAGE 1

Every letter is assigned a value as shown below. Words are assigned values by multiplying the values of their letters. For example, the word *zephyr* has a value of $26 \cdot 5 \cdot 16 \cdot 8 \cdot 25 \cdot 18 = 7,488,000$

A	1	J	10	S	19
B	2	K	11	T	20
C	3	L	12	U	21
D	4	M	13	V	22
E	5	N	14	W	23
F	6	O	15	X	24
G	7	P	16	Y	25
H	8	Q	17	Z	26
I	9	R	18		

Let's find a person's first name with the value 161,000. We'll take the process in steps.

Problem #1

1. Find the prime factorization of 161,000.

Questions and Conversations for #1

This section contains ideas for conversations, mainly in the form of questions that students may ask or that you may pose to them. Be sure to allow students to do most of the thinking and talking!

Teacher's Note. In my experience, students are often excited to begin solving the word puzzles and they may at first be impatient with Problems #1–5. I usually tell them that if they take the time to think carefully about these questions, they will learn much more and have more fun with the activity in the end.

Solution for #1

The prime factorization of 161,000 is $2 \cdot 2 \cdot 2 \cdot 5 \cdot 5 \cdot 5 \cdot 7 \cdot 23$.

Problem #2

2. Why can't the letters C, F, I, L, O, R, U, and X be used? (Try to answer this without testing each letter separately.)

Questions and Conversations for #2

» *How are the letters distributed within the alphabet?* They occur at every third position.
» *What do the prime factorizations of the values of these letters have in common?* They all have a factor of 3.
» *Why is it a problem for them to have a factor of 3?* The number 161,000 does not have a factor of 3.

Teacher's Note for #1. If students do not remember what a prime factorization is, this is a good chance to review the concept. But, instead of reteaching a specific procedure, ask them why, for example, the expression $2 \cdot 2 \cdot 3 \cdot 5$ is the prime factorization of 60 (because the product is 60 and the factors are all prime numbers). Then ask how they could have calculated it themselves. Some students may suggest factor trees, but others may have different approaches. Give them one or two other numbers to practice with. If they make errors, use the opportunity to correct any misconceptions.

Solution for #2

No letter whose value has a factor of 3 can be used because the prime factorization of 161,000 has no factor of 3.

Problem #3

3. Find seven other letters that cannot be used. Explain your thinking.

Questions and Conversations for #3

» *How can you use the prime factorization of 161,000 to answer these questions quickly (without using a calculator)?* Think about comparing the prime factors in 161,000 to the prime factors in the values of the letters.

Solution for #3

K, V, M, Z, Q, and S cannot be used because their values have prime factors of 11, 13, 17, or 19, none of which appear in the prime factorization of 161,000.

The letter P cannot be used because its prime factorization contains four factors of 2 while the prime factorization of 161,000 contains only three factors of 2.

Problem #4

4. Why must the name contain the letter W?

Questions and Conversations for #4

See Questions and Conversations for #3.

Solution for #4

The name must contain W because the prime factorization of 161,000 has a prime factor of 23. W is the only letter that contains this factor.

Problem #5

5. Why must the name contain either G or N, but not both?

Questions and Conversations for #5

See Questions and Conversations for #3.

Solution for #5

The name must have a G or an N because the prime factorization of 161,000 contains the number 7. G, N, and U are the only letters whose values have a prime factor of 7. U cannot be used because its value has a factor of 3. G and N cannot both appear because the number 161,000 has only one factor of 7.

Problem #6

6. Find a person's first name with the value 161,000. Explain your thinking.

Questions and Conversations for #6

» *Which letters must be present? Which letters cannot be present? How many of each letter can you have?* This is where you make use of the information you have gathered so far!

» *What is a quick way to prove your answers are correct without using a calculator?* Check that the combined factors in the values of the letters are identical to the factors in the prime factorization.

Solution for #6

The name WENDY works.

We have seen that the name must contain W and either G or N. It may also have A, B, D, E, H, J, T, or Y. The letter Y seems a reasonable choice because it often appears as the last letter in first names (and nicknames) and it contains two of the three needed factors of 5.

It appears that the name is likely to contain the letters W, N, and Y or W, G, and Y. The first combination might suggest the name WENDY. We can check that this works:

W	E	N	D	Y
23	5	$2 \cdot 7$	$2 \cdot 2$	$5 \cdot 5$

This contains exactly three factors of 2, three factors of 5, one factor of 7, and one factor of 23—exactly the same as the prime factorization of 161,000.

Note: The name DWAYNE will also work!

Problem #7

7. Use prime factorizations to find a word that fits the clue for each value. Prove that your answers are correct.

Value: 83,538	a task at school
Value: 51,129	a good way to exercise
Value: 140,625	a classic toy
Value: 152,460	a person's first name
Value: 43,740	something on a house in winter

Questions and Conversations for #7

See Questions and Conversations for #6.

Solution for #7

The words are QUIZ, SWIM, YOYO, KEVIN, and ICICLE.

$83,538 = 2 \cdot 3 \cdot 3 \cdot 3 \cdot 7 \cdot 13 \cdot 17$

17	$3 \cdot 7$	$3 \cdot 3$	$2 \cdot 13$
Q	U	I	Z

$51,129 = 3 \cdot 3 \cdot 13 \cdot 19 \cdot 23$

19	23	$3 \cdot 3$	13
S	W	I	M

$140,625 = 3 \cdot 3 \cdot 5 \cdot 5 \cdot 5 \cdot 5 \cdot 5 \cdot 5$

$5 \cdot 5$	$3 \cdot 5$	$5 \cdot 5$	$3 \cdot 5$
Y	O	Y	O

$152,460 = 2 \cdot 2 \cdot 3 \cdot 3 \cdot 5 \cdot 7 \cdot 11 \cdot 11$

11	5	$2 \cdot 11$	$3 \cdot 3$	$2 \cdot 7$
K	E	V	I	N

$43,740 = 2 \cdot 2 \cdot 3 \cdot 3 \cdot 3 \cdot 3 \cdot 3 \cdot 3 \cdot 3 \cdot 5$

$3 \cdot 3$	3	$3 \cdot 3$	3	$2 \cdot 2 \cdot 3$	5
I	C	I	C	L	E

STAGE 2

Problem #8

8. Find a number that cannot be the value of any word. Explain your thinking.

Questions and Conversations for #8

» *What types of prime factors cannot occur in the number?* The number cannot have any prime factor greater than 23. Why?

Solution for #8

A number whose prime factorization contains a prime factor larger than 23 will not be the value of any word because 23 is the largest possible prime factor of the value of any letter. For example, the number 620 is not the value of any word because it has a prime factor of 31, which is larger than 23.

Problem #9

9. Many words have values with one or more zeros at the end of the numeral. Why do you think this happens so often?

Questions and Conversations for #9

» *What can you say about the prime factorization of a number whose final digit is 0? What about 00?* You can easily predict how many zeros will be at the end of the numeral just by glancing at the prime factorization. How? (Think about the factors 2 and 5.)

Solution for #9

Many word values have numerals that end in one or more zeros because $10 = 2 \cdot 5$, and 2 and 5 occur frequently as factors of the values of the letters of the alphabet. For every factor pair of 2 and 5, there will be a zero at the end of the numeral.

Problem #10

10. Find the prime factorization of 720. Use it to find at least six words that have this value. Use the prime factorization to show that each word is correct.

Solution for #10

The prime factorization of 720 is $2 \cdot 2 \cdot 2 \cdot 2 \cdot 3 \cdot 3 \cdot 5$. Here are some possibilities:

COP	$3 \cdot (3 \cdot 5) \cdot (2 \cdot 2 \cdot 2 \cdot 2)$	PIE	$(2 \cdot 2 \cdot 2 \cdot 2) \cdot (3 \cdot 3) \cdot 5$
HER	$(2 \cdot 2 \cdot 2) \cdot 5 \cdot (2 \cdot 3 \cdot 3)$	HEAR	$(2 \cdot 2 \cdot 2) \cdot 5 \cdot 1 \cdot (2 \cdot 3 \cdot 3)$
DIED	$(2 \cdot 2) \cdot (3 \cdot 3) \cdot 5 \cdot (2 \cdot 2)$	LACED	$(2 \cdot 2 \cdot 3) \cdot 1 \cdot 3 \cdot 5 \cdot (2 \cdot 2)$
BRAT	$2 \cdot (2 \cdot 3 \cdot 3) \cdot 1 \cdot (2 \cdot 2 \cdot 5)$	CHAFE	$3 \cdot (2 \cdot 2 \cdot 2) \cdot 1 \cdot (2 \cdot 3) \cdot 5$
BLOB	$2 \cdot (2 \cdot 2 \cdot 3) \cdot (3 \cdot 5) \cdot 2$	TALC	$(2 \cdot 2 \cdot 5) \cdot 1 \cdot (2 \cdot 2 \cdot 3) \cdot 3$

Each of the products has exactly four factors of 2; two factors of 3; and one factor of 5—the same as the prime factorization of 720.

Problem #11

11. Suppose each letter is assigned a unique prime number value:

A	2	J	29	S	67
B	3	K	31	T	71
C	5	L	37	U	73
D	7	M	41	V	79
E	11	N	43	W	83
F	13	O	47	X	89
G	17	P	53	Y	97
H	19	Q	59	Z	101
I	23	R	61		

Do you think this makes the puzzle easier or harder? In what other interesting ways does it change the puzzle? Give examples.

Questions and Conversations for #11

> » *What's the best way to get a feel for how the new system works?* The best thing is probably just to try it. Choose some values and search for words for them.
>
> » *What will happen if you are asked to find as many words as possible with a certain value? (See Question 10.) Why does this happen?* Again, if you're not sure, then try it!

Solution for #11

It makes the puzzle harder in one way, because the values will be larger. For example, WENDY will now have a value of 26,656,861 (try it)! It will definitely take longer to find the prime factorizations of these numbers.

On the other hand, once you find the prime factorization, it will be easier than before because you will always know exactly which letters will be in the word. This is because there is no way to create a prime value using values of other letters!

For example, the value 11,011 has a prime factorization of $7 \cdot 11 \cdot 11 \cdot 13$. This tells you that the word will have a D, two E's, and an F. The only common word fitting this description is FEED. When you use prime numbers as values, if two words have the same value, they must be anagrams!

STAGE 3

Problem #12

12. Show how to use the prime factorization of 4998 to find all of its factors. How can you tell when you have found them all?

Questions and Conversations for #12

> » *What have you learned about the connections between factors and prime factorizations in this exploration? How can this help you?* All factors of a number are just products of some combination of the numbers in its prime factorization. The key is to find all of these combinations!
>
> » *How can you organize your work effectively?* There are many ways. Might it help to think of factors in pairs? Or to keep track of how many prime factors each number has?
>
> » *What number does it represent when all the prime factors are used? What about when none of them are used?* When all of the prime factors are present, it stands for the original number, 4998. When none of them are present, it stands for the number 1! Can you see why?

Solution for #12

The prime factorization of 4998 is $2 \cdot 3 \cdot 7 \cdot 7 \cdot 17$. The factors of 4998 are all of the products that can be created from these five prime numbers.

If we organize our work carefully, we can make sure to find them all. We begin by creating pairs built from one and four prime factors:

$$2 \cdot (3 \cdot 7 \cdot 7 \cdot 17) = 2 \cdot 2499 \qquad 3 \cdot (2 \cdot 7 \cdot 7 \cdot 17) = 3 \cdot 1666$$

$$7 \cdot (2 \cdot 3 \cdot 7 \cdot 17) = 7 \cdot 714 \qquad 17 \cdot (2 \cdot 3 \cdot 7 \cdot 7) = 17 \cdot 294$$

Now we group by twos and threes:

$$(2 \cdot 3) \cdot (7 \cdot 7 \cdot 17) = 6 \cdot 833 \qquad (2 \cdot 7) \cdot (3 \cdot 7 \cdot 17) = 14 \cdot 357$$

$$(2 \cdot 17) \cdot (3 \cdot 7 \cdot 7) = 34 \cdot 147 \qquad (3 \cdot 7) \cdot (2 \cdot 7 \cdot 17) = 21 \cdot 238$$

$$(3 \cdot 17) \cdot (2 \cdot 7 \cdot 7) = 51 \cdot 98 \qquad (7 \cdot 7) \cdot (2 \cdot 3 \cdot 17) = 49 \cdot 102$$

$$(7 \cdot 17) \cdot (2 \cdot 3 \cdot 7) = 119 \cdot 42$$

The only remaining combination is:

$$1 \cdot (2 \cdot 3 \cdot 7 \cdot 7 \cdot 17) = 1 \cdot 4998$$

There are 24 factors in all:

$$1, 2, 3, 6, 7, 14, 17, 21, 34, 42, 49, 51, 98, 102, 119, 147,$$
$$238, 294, 357, 714, 833, 1666, 2499, \text{ and } 4998$$

WRAP UP

Share Strategies

Have students share their strategies and compare results.

Summarize

Answer any remaining questions that students have.

Summarize the key concept of the exploration—every factor of a number (except 1) is a product of some or all of its prime factors. If students completed Stage 3, they used this idea to find all of the factors of 4998. You could reinforce this idea by giving them a few more numbers to try this with.

Further Exploration

Ask students to think of ways to continue or extend this exploration. Here are some possibilities:

- » Create and share Factor Scramble puzzles with classmates. Try other systems for assigning values to the letters.
- » Create and share crossword puzzles in which the clues are the values of the words.
- » What types of numbers have many words associated with them? (See Problem #10.) Try to find a value that has as many words as possible.

Exploration 5

How Many Factors?

INTRODUCTION

Materials

» Calculator
» Completed "Building Blocks Grid" (see http://www.routledge.com/Assets/ClientPages/Advanced CCMath.aspx)

Prior Knowledge

» Complete Stage 1 of Exploration 1: Building Blocks.
» Understand and use natural number exponents.

Learning Goals

» Understand the relationship between a number's factors and its prime factorization.
» Develop a formula to count the factors of a number.
» Apply a formula to solve problems about properties of exponents and factors.
» Analyze connections between prime factorizations, exponents, and factors.
» Communicate complex mathematical ideas clearly.
» Persist in solving challenging problems.

Launching the Exploration

Motivation and purpose. To students: Prime factorizations are powerful tools! You can use them to simplify calculations and to deepen your understanding of many mathematical concepts. In this activity, you will explore how to use them to predict how many factors a number will have. In the process, you will also learn some new things about exponents.

Understanding the problem. Practice finding prime factorizations of a few numbers and showing them as block diagrams (see Exploration 1: Building Blocks). Then write the prime factorizations using exponents. Guide students to notice that each base's exponent is equal to the number of times the base appears in the block diagram. In particular, if a number appears once in the prime factorization, you may write it with an exponent of 1.

DOI: 10.4324/9781003232742-7

Suggest that students look at the "Building Blocks" grid, find out how many factors some of the numbers have, compare this to their block diagrams, and look for patterns. Have them share and record any interesting observations for later reference. (For example, students might notice that numbers with two vertically stacked blocks always have four factors, or that numbers with three blocks in an "L" shape always have six factors.)

Note. Instead of coloring the blocks in this activity, we will just write the prime number inside of each block. We will also choose to arrange the blocks in the way that is the most convenient. (Sometimes, it's easiest just to write them horizontally.)

STUDENT HANDOUT

Stage 1

1. List the factors of 16 and write a block diagram and an exponential expression for each. How many factors are there? How could you have predicted this from the exponential expression for 16? Explain.

2. Write a block diagram and an exponential expression for each factor of 20. How many factors does 20 have? How does this relate to the number of twos and fives in its prime factorization? Explain.

3. Write a block diagram and an exponential expression for each factor of 72. How many factors does 72 have? How could you have predicted this from the exponential expression for 72? Explain.

4. Find the prime factorization of 567. Use the prime factorization (in exponential form) to predict the number of factors of 567. Explain your thinking. Check your prediction by finding all of the factors.

5. Describe a general method for predicting how many factors a number has from its prime factorization. Explain why your method works.

6. Choose a number that has three different prime factors. Adjust and apply your method to this number. Does it work?

Stage 2

7. Suppose the prime factorization of some number is $a^x \cdot b^y \cdot c^z$. How many factors (N) does the number have? Explain. We will call your answer the Number of Factors Formula.

8. What is the smallest number that has exactly 18 factors? The next smallest? Show how to use the Number of Factors Formula to help you answer these questions.

9. Square numbers have an odd number of factors. (If you completed Exploration 2: 1000 Lockers, you learned about this!) Use this fact and the Number of Factors Formula to explain why a square number must have an even number of each prime factor in its prime factorization.

Stage 3

10. Choose at least four natural numbers that are not square numbers. For each number, gather this data:
    - » the number you chose (x)
    - » how many factors it has (y)
    - » the product of all of its factors (z)

Discover a formula that relates x, y, and z. Explain why the formula works. Why were you asked not to choose a square number?

TEACHER'S GUIDE

STAGE 1

Problem #1

1. List the factors of 16 and write a block diagram and an exponential expression for each. How many factors are there? How could you have predicted this from the exponential expression for 16? Explain.

Questions and Conversations for #1

This section contains ideas for conversations, mainly in the form of questions that students may ask or that you may pose to them. Be sure to allow students to do most of the thinking and talking!

» *What does each exponent tell you about the number of blocks (prime factors) in the block diagram?* It tells you how many blocks there are for that factor.

» *How can you show the factor 1 using exponential notation?* You can write it as 2^0. The exponent of 0 shows that there are no blocks (no factors of 2).

» *How do the exponents on the factors of 16 compare to the exponent for 16 itself?* The exponent for 16 is 4. The exponents of the factors are 0, 1, 2, 3, and 4. The exponents of the factors are all less than or equal to the exponent for 16.

» *How does the exponent for 16 compare with the number of factors?* The exponent for 16 is 4. There are 5 factors. The number of factors is one greater than the exponent. (What causes this?)

Solution for #1

Each exponent matches the number of 2-blocks (factors of 2). Every time you divide by 2, the number of blocks and the exponent each decrease by 1. By the time you reach 1, there are no 2-blocks left, so the exponential expression is 2^0!

The number 16 has five factors, one for each exponent less than or equal to 4, including 0. This makes the number of factors one greater than the exponent in 2^4.

16	8	4	2	1
2 2 2 2	2 2 2	2 2	2	
2^4	2^3	2^2	2^1	2^0

Problem #2

2. Write a block diagram and an exponential expression for each factor of 20. How many factors does 20 have? How does this relate to the number of twos and fives in its prime factorization? Explain.

Questions and Conversations for #2

» *How will you show an exponential expression with a base of 2 when there are no 2-blocks in the factor?* You can write it as 2^0. (This also applies to other bases.)

» *How many 2-blocks can be in a factor of 20? List the possibilities. How many 5-blocks?* There can be 0, 1, or 2 of the 2-blocks in a factor of 20. There can be 0 or 1 5-blocks.

» *Can you think of a way to organize the factors of 20 according to how many twos and fives are present in each factor?* There are many ways to do this, including tables and tree diagrams. Another method is shown in the solution to this question. (Be sure to allow students plenty of time to practice creating their own methods to organize the information before stepping in to assist.)

Solution for #2

The number 20 has six factors.

20	10	5	4	2	1
2 2 5	2 5	5	2 2	2	
$2^2 \cdot 5^1$	$2^1 \cdot 5^1$	$2^0 \cdot 5^1$	$2^2 \cdot 5^0$	$2^1 \cdot 5^0$	$2^0 \cdot 5^0$

By looking closely at the prime factorizations, you can see that each factor of 20 has 2, 1, or 0 twos, and 1 or 0 fives. Organizing this information in other ways might make it easier to see why there must be six factors of 20. Here is one way to do it:

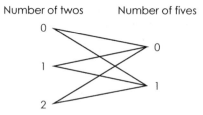

Number of twos   Number of fives

Each of the six line segments represents a factor. For example, the top segment connecting 0 to 0 stands for $2^0 \cdot 5^0$, which equals 1. The next one connecting 0 on the left to 1 on the right represents $2^0 \cdot 5^1$, which equals 5, etc. Since each of the 3 numbers in the left column is connected to 2 numbers in the right column, there are $3 \cdot 2 = 6$ factors of 20.

Problem #3

3. Write a block diagram and an exponential expression for each factor of 72. How many factors does 72 have? How could you have predicted this from the exponential expression for 72? Explain.

Solution for #3

Students may organize their work in various ways, including the method shown in the solution to Problem #2. Here, we'll show two other methods that also work well—especially when numbers have many factors.

> ***Teacher's Note for #3.*** You can ask the same types of questions as in #2. However, because there are now more factors, some students might want to refine their systems of organizing them. This might be a good time to suggest tables or tree diagrams if they haven't thought of them already.

Sample student response 1 (table):

Number of threes \ Number of twos	0	1	2	3
0	$2^0 \cdot 3^0 = 1$	$2^1 \cdot 3^0 = 2$ [2]	$2^2 \cdot 3^0 = 4$ [2][2]	$2^3 \cdot 3^0 = 8$ [2][2][2]
1	$2^0 \cdot 3^1 = 3$ [3]	$2^1 \cdot 3^1 = 6$ [3] [2]	$2^2 \cdot 3^1 = 12$ [3] [2][2]	$2^3 \cdot 3^1 = 24$ [3] [2][2][2]
2	$2^0 \cdot 3^2 = 9$ [3][3]	$2^1 \cdot 3^2 = 18$ [3][3] [2]	$2^2 \cdot 3^2 = 36$ [3][3] [2][2]	$2^3 \cdot 3^2 = 72$ [3][3] [2][2][2]

Sample student response 2 (tree diagram):

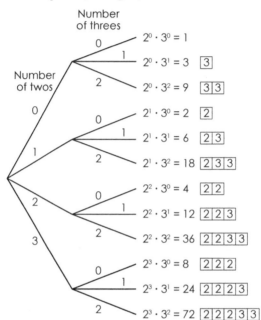

The number 72 has 12 factors. The table (or the tree diagram) shows that you can find this by multiplying $4 \cdot 3$. This works because there are 4 possible values for the number of twos (0, 1, 2, or 3), and 3 possible values for the number of threes (0, 1, or 2).

Because 0 is always one of the possible values, the number of values is always one greater than the exponent. This means that you can just add 1 to each exponent in the prime factorization of 72, and then multiply the results!

$$72 = 2^{\boxed{3}} \cdot 3^{\boxed{2}}$$
$$(\boxed{3}+1) \cdot (\boxed{2}+1) = 4 \cdot 3 = 12$$

Problem #4

4. Find the prime factorization of 567. Use the prime factorization (in exponential form) to predict the number of factors of 567. Explain your thinking. Check your prediction by finding all of the factors.

Solution for #4

$567 = 3^4 \cdot 7^1$

There are 10 factors. You can find this by adding one to each exponent and multiplying the results:

$(4+1) \cdot (1+1) = 5 \cdot 2 = 10$

One way to check this prediction is to make a table:

Teacher's Note for #4. This item requires students to test, generalize, and apply what they have discovered so far. If they are struggling, suggest that they experiment with more numbers (27, 25, and 15 might be good choices). Continue to pose questions that help them focus on organization strategies and patterns.

Factors of 567

Number of threes

		0	1	2	3	4
Number of sevens	0	$3^0 \cdot 7^0 = 1$	$3^1 \cdot 7^0 = 3$ $\boxed{3}$	$3^2 \cdot 7^0 = 9$ $\boxed{3}\boxed{3}$	$3^3 \cdot 7^0 = 27$ $\boxed{3}\boxed{3}\boxed{3}$	$3^4 \cdot 7^0 = 81$ $\boxed{3}\boxed{3}\boxed{3}\boxed{3}$
	1	$3^0 \cdot 7^1 = 7$ $\boxed{7}$	$3^1 \cdot 7^1 = 21$ $\boxed{7}$ $\boxed{3}$	$3^2 \cdot 7^1 = 63$ $\boxed{7}$ $\boxed{3}\boxed{3}$	$3^3 \cdot 7^1 = 189$ $\boxed{7}$ $\boxed{3}\boxed{3}\boxed{3}$	$3^4 \cdot 7^1 = 567$ $\boxed{7}$ $\boxed{3}\boxed{3}\boxed{3}\boxed{3}$

There are five columns and two rows, for a total of 10 factors.

You can also show this with a tree diagram having five branches, each splitting into two (or vice versa).

Problem #5

5. Describe a general method for predicting how many factors a number has from its prime factorization. Explain why your method works.

Questions and Conversations for #5

See Teacher's Note for #4.

Solution for #5

To find how many factors a number has, add 1 to each exponent in its prime factorization and multiply the results.

This works because when a prime factor has an exponent of n, there are $n + 1$ possible values for the number of times it can appear in a factor of the number. The "plus 1" is in the formula because a factor may appear 0 times. You multiply to find the total number of combinations.

Problem #6

6. Choose a number that has three different prime factors. Adjust and apply your method to this number. Does it work?

Teacher's Note. Students who decide to make a table will have to think of a way to represent a third dimension in their diagram. This may require some creativity!

Questions and Conversations for #6

See Teacher's Note for #4.

Solution for #6

Sample student response:

$$60 = 2^2 \cdot 3^1 \cdot 5^1 \qquad \text{Number of factors: } (2+1) \cdot (1+1) \cdot (1+1) = 3 \cdot 2 \cdot 2 = 12$$

The 12 factors are 1, 2, 3, 4, 5, 6, 10, 12, 15, 20, 30, and 60. You could also show this in a tree diagram:

Factors of 60

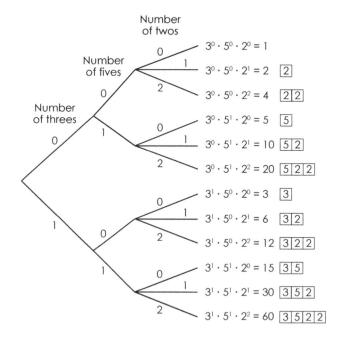

STAGE 2

Problem #7

7. Suppose the prime factorization of some number is $a^x \cdot b^y \cdot c^z$. How many factors (N) does the number have? Explain. We will call your answer the Number of Factors Formula.

Questions and Conversations for #7

» *Can you create the formula by using your process for predicting the number of factors?* Yes! People often create formulas by finding a process first. In fact, the algebraic formula is just a different way to describe the process.

» *Does the formula take into account the order of operations?* It should! Are parentheses needed? Why or why not?

» *This version of the formula applies when there are three different prime factors. Can you create versions that work when there are more (or fewer) prime factors?* Yes. Try showing some of them!

» *Does the formula still work if some (or all) of the exponents are 0?* Yes. For example, we can write the number 4 as:

$$4 = 2^2 \cdot 3^0$$

The formula tells you that 4 has three factors:

$$(2+1) \cdot (0+1) = 3 \cdot 1 = 3$$

Including 3^0 in the exponential expression did not affect the answer—which is a good thing because 3 is not a factor of 4!

Solution for #7

$N = (x+1) \cdot (y+1) \cdot (z+1)$. This formula just shows what we described before in words: Add one to each exponent in the prime factorization and multiply the results.

Problem #8

8. What is the smallest number that has exactly 18 factors? The next smallest? Show how to use the Number of Factors Formula to help you answer these questions.

Questions and Conversations for #8

» *What numbers can you multiply to get 18?* Try listing the possibilities. Don't forget that you can multiply more than two numbers!

> » *Can the factors of 18 tell you anything about the exponents in the prime factorizations of the unknown numbers?* Yes. Your formula can show you how.
> » *How do you know what bases to use in the prime factorizations of the unknown numbers?* Experiment! How can you make the numbers as small as possible?

Solution for #8

The smallest number that has 18 factors is 180. The next smallest is 252.

We can write 18 as $18 \cdot 1$, $9 \cdot 2$, $6 \cdot 3$, or $3 \cdot 3 \cdot 2$

The exponents in the prime factorizations will be one less than these numbers. The possibilities are:

| 17 and 0 | 8 and 1 | 5 and 2 | 2, 2, and 1 |

To make the answers as small as possible, we use the smallest bases we can. We also put the smaller bases with the larger exponents. This gives four possibilities so far:

$$2^{17} = 131{,}072 \qquad 2^8 \cdot 3^1 = 768 \qquad 2^5 \cdot 3^2 = 288 \qquad 2^2 \cdot 3^2 \cdot 5^1 = 180$$

We should also try the next smallest set of bases for the last one.

$$2^2 \cdot 3^2 \cdot 7^1 = 252$$

Because 252 is less than 288, the two smallest answers are 180 and 252.

Problem #9

9. Square numbers have an odd number of factors. (If you completed Exploration 2: 1000 Lockers, you learned about this!) Use this fact and the Number of Factors Formula to explain why a square number must have an even number of each prime factor in its prime factorization.

Questions and Conversations for #9

> » Do you know what happens when you multiply different combinations of odd and even numbers?
>
> | even $\cdot$ even = ? | even $\cdot$ odd = ? |
> | odd $\cdot$ even = ? | odd $\cdot$ odd = ? |

If you don't, then experiment and find out! Try to understand what makes the answers come out the way they do.

Solution for #9

We need to get an odd answer from the Number of Factors Formula. In order for a product to be odd, each factor must be odd. Because each factor in the formula is one greater than the corresponding exponent in the prime factorization, the exponents must be even numbers. This means there is an even number of each prime factor.

STAGE 3

Problem #10

10. Choose at least four natural numbers that are not square numbers. For each number, gather this data:
    » the number you chose (x)
    » how many factors it has (y)
    » the product of all of its factors (z)

Discover a formula that relates x, y, and z. Explain why the formula works. Why were you asked not to choose a square number?

Questions and Conversations for #10

» *How can you find a pattern when the products get so large?* This can be a challenge. Consider choosing very small numbers (less than 20).
» *How can you organize the data that you collect? How might this help?* One method is to make a table. This might make it easier to find patterns.
» *What can you do to x and y to get z?* This is the question that will lead you to your formula. What observations can you make about the values of x, y, and z? Does the fact that the z values are often so large give you any clues?
» *Is there a helpful way to reorder or regroup the numbers when you are finding the product of all the factors?* Yes, there is. Explore!

Solution for #10

Sample student response: I chose the numbers 6, 7, 10, and 12. Then I made a table to show x, y, and z.

x (chosen number)	6	7	10	12
y (number of factors)	4	2	4	6
z (product of the factors)	36	7	100	1,728

$$z = x^{\frac{y}{2}}$$

Group together the pairs of factors that equal x. Because there are half as many pairs as factors (y), the product of all of them is $x^{\frac{y}{2}}$. For example, the factors of 12 are 1, 2, 3, 4, 6, and 12. When you multiply them, you can change the order and regroup them to get

$$1 \cdot 2 \cdot 3 \cdot 4 \cdot 6 \cdot 12 = (1 \cdot 12) \cdot (2 \cdot 6) \cdot (3 \cdot 4) = 12 \cdot 12 \cdot 12 = 12^3 = 1728$$

The exponent, 3, is half the number of factors because the 6 factors are grouped into 3 pairs.

If x is a square number, it has an odd number of factors (y). When you divide this by 2 to get the exponent in the formula, it is not a whole number!

WRAP UP

Share Strategies

Ask students to share and compare strategies and observations. Help them make connections between tables, tree diagrams, and any other representations they used to organize their data.

Summarize

Answer remaining questions that students have. Summarize these key points:

» Every exponent in the prime factorization of any factor of a number N is less than or equal to the corresponding exponent in the prime factorization of N itself.

» Exponents count the number of blocks (prime factors) in a block diagram. You can show "no blocks" as an exponent of zero. This suggests that if p is any prime number, $p^0 = 1$. (This is also true if p is not prime, unless it is equal to 0).

» To find the number of factors of N, add 1 to each exponent in its prime factorization and multiply the results. The reason for adding 1 is to account for the fact that a prime factor may appear zero times.

Further Exploration

Have students think of new questions to ask or ways to continue or extend this exploration. Here are some possibilities:

» Return to the "Building Blocks" grid and think again about how many factors the numbers on the grid have. Do you see any new patterns? If you saw patterns before, do you understand them better now?

» Try some extreme cases of Problem #8. For example, find the smallest number that has 100 factors.

» If you know how many factors a and b have, can you predict the number of factors that $a \cdot b$ has?

» The question in Stage 3 might make you wonder about the meaning of non-whole-number exponents. Can you suggest any possible conjectures about this? Can you think of evidence to support or disprove your conjectures?

Exploration 6

Differences and Greatest Common Factors

INTRODUCTION

Prior Knowledge

» Understand the relationship between factors and multiples.
» Calculate greatest common factors using lists of factors.
» Simplify fractions.
» Calculate least common multiples using lists of multiples (Stage 3 only).

Learning Goals

» Increase fluency with calculating greatest common factors.
» Discover a connection between differences and greatest common factors.
» Analyze and extend complex patterns.
» Apply knowledge of the difference-GCF connection to simplify fractions.
» Communicate complex mathematical ideas clearly.
» Persist in solving challenging problems.

Launching the Exploration

Motivation and purpose. To students: How would you go about simplifying a fraction like $\frac{6538}{6539}$? In this exploration, you will unearth an interesting connection between factors and subtraction that will make this much easier than you might expect!

Understanding the problem. Check for prior knowledge by asking students to find the greatest common factor of one or two pairs of numbers.

Discuss the vocabulary in the exploration:

» The terms "*difference*" and "*differ by*" involve the idea of subtraction.
» The *natural numbers* are the counting numbers: $1, 2, 3, 4, 5, \ldots$ (In this exploration, all differences are assumed to be natural numbers.)
» A *conjecture* is an informed guess or hypothesis. Mathematicians spend much of their time making, testing, and proving or disproving conjectures.

Briefly look through the exploration with students. Point out that Problem #6 is the main goal of the activity—to discover and describe a general connection between differences and greatest common factors. The earlier problems guide them toward this goal step by step. Students who work on Stages 2 and 3 will have a chance to apply and extend their new knowledge.

DOI: 10.4324/9781003232742-8

STUDENT HANDOUT

Stage 1

1. Show how to find the greatest common factor (GCF) of each pair. Then make a conjecture about greatest common factors of natural numbers that differ by 1.

   (a) 7, 8   (b) 10, 11   (c) 14, 15   (d) 25, 24   (e) 35, 36

2. Test your conjecture on the following pairs. Show your work.

   (a) 80, 81   (b) 135, 136

3. Experiment with greatest common factors of pairs that have differences of 2 and 3. Make and test a conjecture for each case.

4. Without doing any more calculations, use your experience so far to make a conjecture about greatest common factors of pairs that differ by 4. Explain your thinking.

5. Test your prediction. Were you correct? If not, explain what went wrong.

6. What are the possible greatest common factors of numbers that differ by n? Answer the question by continuing to explore GCFs for greater differences. Describe your thinking process.

7. Explain how you can be certain that the fractions $\frac{5874}{5875}$ and $\frac{2947}{2949}$ are in simplest form without finding factors of the numerators and denominators.

8. Are the fractions $\frac{3465}{3479}$ and $\frac{555}{596}$ in simplest form? Explain how to use your knowledge of differences and GCFs to find the answers as easily as possible. (See if you can limit the use of your calculator to one computation for each fraction.)

9. You may have noticed that if you arrange pairs having the same difference into organized lists, patterns appear in the GCFs. For what types of differences are the patterns the simplest? The most complex? Explain.

10. Create a general method to predict the exact pattern of greatest common factors for any difference (without actually calculating any of the GCFs). Illustrate your method on pairs that have a difference of 12.

Stage 3

11. Search for a connection between differences and least common multiples. Describe any discoveries that you make.

TEACHER'S GUIDE

STAGE 1

Problem #1

1. Show how to find the greatest common factor (GCF) of each pair. Then make a conjecture about greatest common factors of natural numbers that differ by 1.

(a) 7, 8   (b) 10, 11   (c) 14, 15   (d) 25, 24   (e) 35, 36

Solution for #1

7: 1, 7	8: 1, 2, 4, 8	GCF(7, 8) = 1
10: 1, 2, 5, 10	11: 1, 11	GCF(10, 11) = 1
14: 1, 2, 7, 14	15: 1, 3, 5, 15	GCF(14, 15) = 1
25: 1, 5, 25	24: 1, 2, 3, 4, 6, 8, 12, 24	GCF(25, 24) = 1
35: 1, 5, 7, 35	36: 1, 2, 3, 4, 6, 9, 12, 18, 36	GCF(35, 36) = 1

Conjecture: Natural numbers that differ by 1 always have a greatest common factor of 1.

Problem #2

2. Test your conjecture on the following pairs. Show your work.
    (a) 80, 81   (b) 135, 136

Solution for #2

80: 1, 2, 4, 5, 8, 10, 16, 20, 40, 80	81: 1, 3, 9, 27, 81	GCF(80, 81) = 1
135: 1, 3, 5, 9, 15, 27, 45, 135	136: 1, 2, 4, 8, 17, 34, 68, 136	GCF(135, 136) = 1

Both results support the conjecture.

Problem #3

3. Experiment with greatest common factors of pairs that have differences of 2 and 3. Make and test a conjecture for each case.

Questions and Conversations for #3

This section contains ideas for conversations, mainly in the form of questions that students may ask or that you may pose to them. Be sure to allow students to do most of the thinking and talking!

» *How can you be confident that you've found all possible greatest common factors?* The more pairs you try, of course, the more confident you can be! Another approach is to arrange the pairs into organized lists and look for patterns.

» *How can you be certain that you've found all possible greatest common factors?* Notice the difference between confidence and certainty! Trying many pairs and looking for patterns can increase your confidence, but the only way to be absolutely certain you've found all the possibilities is to understand what actually causes the patterns.

» *How do you know what numbers to use when you test the conjectures?* There's no simple answer. Aim for variety. For example, if you've tested many small numbers, try some larger ones. If you've tested numbers that don't have many factors, try some that have a lot of factors.

Solution for #3

Sample student response 1: Here are a few examples.

Difference of 2: GCF(4, 6) = 2   GCF(10, 12) = 2   GCF(7, 9) = 1
Difference of 3: GCF(2, 5) = 1   GCF(10, 13) = 1   GCF(6, 9) = 3

After trying many more pairs, you can make these conjectures:
» If the difference of two numbers is 2, their GCF is always 1 or 2.
» If the difference of two numbers is 3, their GCF is always 1 or 3.

To test the conjectures, you can choose some larger numbers. Here are two examples with differences of 2.

143: 1, 11, 13, 143   145: 1, 5, 29, 145   GCF(143, 145) = 1

198: 1, 2, 3, 6, 9, 11, 18, 22,   200: 1, 2, 4, 5, 8, 10, 20,   GCF(198, 200) = 2
33, 66, 99, 198   25, 40, 50, 100, 200

The conjectures still appear to be true.

Sample student response 2: You can make tables to keep track of the differences and greatest common factors.

Differences of 2		Differences of 3	
Number Pair	GCF	Number Pair	GCF
1, 3	1	1, 4	1
2, 4	2	2, 5	1
3, 5	1	3, 6	3
4, 6	2	4, 7	1
5, 7	1	5, 8	1
6, 8	2	6, 9	3

Based on the patterns, you might make these conjectures about what happens when you organize the pairs in a list:

» When the difference is 2, the GCF will always alternate between 1 and 2.
» When the difference is 3, the GCF will always be 1 or 3 and every third GCF will be 3.

To test the conjecture, extend the tables to check that the patterns continue.

Problem #4

4. Without doing any more calculations, use your experience so far to make a conjecture about greatest common factors of pairs that differ by 4. Explain your thinking.

Questions and Conversations for #4

» *What if your conjecture is wrong?* Don't worry. Mathematicians make wrong conjectures all the time. It can be fun to be surprised when you discover the truth! You often learn just as much—or more—from an incorrect conjecture as a correct one. You should have a good reason for your prediction, though.

Solution for #4

Sample student response: Conjecture: The GCF of numbers that differ by 4 will always be either 1 or 4. Reason: This seems sensible because so far, every GCF has been either 1 or the difference between the numbers, so this might happen here, too.

Problem #5

5. Test your prediction. Were you correct? If not, explain what went wrong.

Solution for #5

Sample student response: The conjecture was not correct. There were many pairs that had greatest common factors of 1 and 4, but other pairs had a greatest common factor of 2, such as:

$$GCF(2, 6) = 2 \quad GCF(18, 22) = 2$$

Problem #6

6. What are the possible greatest common factors of numbers that differ by n? Answer the question by continuing to explore GCFs for greater differences. Describe your thinking process.

Teacher's Note for #5. Some students will miss the GCF of 2, which is reasonable given their experience with differences of 2 and 3. Testing the conjecture gives them a chance to notice this. Encourage them again to test plenty of pairs.

Teacher's Notes. Some students may make additional observations. If they made an organized list, they may have seen a repeating pattern in the GCFs: 1, 2, 1, 4, 1, 2, 1, 4, . . .

Others may suspect (correctly!) that the number 2 became a part of the pattern because it is a factor of the difference, 4.

Questions and Conversations for #6

» *What does it mean to "differ by n"?* Because the variable n can represent any natural number, the answer must be a rule, not a number. You are being asked to discover a rule or pattern for the possible GCFs for any value of the difference.

» *What is the best way to keep track of the data for GCFs and differences? Why is this helpful?* The easiest and clearest way is probably to make a table showing the difference in one column and the list of the GCFs in the other. This may make it easier to spot patterns and make connections.

» *How many differences should you test?* Continue until you are confident that you can predict what will happen for any difference.

» *What causes the patterns between the differences and the GCFs?* This question might be asked by or posed to anyone who has already begun to notice that the GCFs are always factors of the difference.

To explore what causes this, suppose we make a picture. Because the difference between two numbers tells you how far apart they are, our picture could be a number line that uses a bar to show a fixed difference, such as 6:

1 2 3 4 5 **6** 7 8 9 10 11 **12** 13 14 15 16 17 **18**

We've shown multiples of 6 in bold. Because you know that 2 and 3 are also possible GCFs, you could mark them in some way, too. For example, circle all multiples of 2 and put a box around the multiples of 3.

Right now the bar shows the pair 1, 7. Imagine sliding it to the right, so that it shows other pairs with a difference of 6 such as 2, 8 and 3, 9. What happens when the number on the left side of the bar has a factor of 2? 3? 6? What happens when it does not?

Teacher's Note. If students have done Exploration 1 (Building Blocks), they might like to track these differences on the grid instead of a number line, focusing on the patterns in the colored blocks.

Solution for #6

If students make tables, they might look something like this:

Differences of 5	
Number Pair	GCF
1, 6	1
2, 7	1
3, 8	1
4, 9	1
5, 10	5
6, 11	1
7, 12	1
8, 13	1
9, 14	1
10, 15	5
11, 16	1
12, 17	1

Differences of 6	
Number Pair	GCF
1, 7	1
2, 8	2
3, 9	3
4, 10	2
5, 11	1
6, 12	6
7, 13	1
8, 14	2
9, 15	3
10, 16	2
11, 17	1
12, 18	6

They should continue testing pairs with larger differences until they can conclude that if two numbers differ by n, their greatest common factor is always a factor of n.

STAGE 2

Problem #7

7. Explain how you can be certain that the fractions $\dfrac{5874}{5875}$ and $\dfrac{2947}{2949}$ are in simplest form without finding factors of the numerators and denominators.

EXPLORATION 6: DIFFERENCES AND GREATEST COMMON FACTORS

Questions and Conversations for #7

> » *What do greatest common factors have to do with the process of simplifying fractions?* If you're not sure, choose a few fractions to simplify, find the GCF of the numerator and denominator, and look for connections.
> » *How can you apply your new knowledge of the connections between differences and GCFs?* Think about the difference between each numerator and denominator.

Solution for #7

Because 5874 and 5875 have a difference of 1, their greatest common factor is 1. This means that there is no whole number except 1 by which they can both be divided evenly. Therefore, the fraction is in simplest form.

Because 2947 and 2949 have a difference of 2, their only possible common factors are 1 and 2. Because they are not even, 2 is not a factor of either number. Therefore, their greatest common factor is 1. Again, this means that the fraction is in simplest form.

Problem #8

8. Are the fractions $\frac{3465}{3479}$ and $\frac{555}{596}$ in simplest form? Explain how to use your knowledge of differences and GCFs to find the answers as easily as possible. (See if you can limit the use of your calculator to one computation for each fraction.)

Questions and Conversations for #8

See the Questions and Conversations for #7.

Solution for #8

The fraction $\frac{3465}{3479}$ is not in simplest form. 3479 and 3465 have a difference of 14, so their possible common factors are 1, 2, 7, and 14. However, 2 and 14 cannot be factors of odd numbers. Thus, 7 is the only number we have to test. We can calculate that $3465 \div 7$ is a whole number, 495. Because the denominator is 14 larger than the numerator, it must also be divisible by 7, with a quotient of 497. (The fraction can be simplified to $\frac{495}{497}$.)

The fraction $\frac{555}{596}$ is in simplest form—555 and 596 differ by 41, which is a prime number. Therefore, the only possible common factors of 555 and 596 are 1 and 41. You can calculate that 555 (or 596) divided by 41 is not a whole number. Therefore, the greatest common factor must be 1. This means that the fraction is in simplest form.

Problem #9

9. You may have noticed that if you arrange pairs having the same difference into organized lists, patterns appear in the GCFs. For what types of differences are the patterns the simplest? The most complex? Explain.

Questions and Conversations for #9

» *What does it mean to arrange pairs into organized lists?* A natural choice is to put the pairs in numerical order. For example, you could begin them like this:

Differences of 2	
Number Pair	GCF
1, 3	1
2, 4	2
3, 5	1
4, 6	2

Differences of 3	
Number Pair	GCF
1, 4	1
2, 5	1
3, 6	3
4, 7	1

» *What causes the patterns to be more complex?* Numbers with more factors have more complex patterns.

Solution for #9

Prime number differences show the simplest patterns. The pattern for a prime number (p) is just a list of ones interrupted by the prime number p at intervals of p. For example, the pattern for a difference of 7 is

$$1, 1, 1, 1, 1, 1, 7, 1, 1, 1, 1, 1, 1, 7, \ldots$$

The reason that the patterns for prime differences are so simple is that prime numbers have only two factors. Our investigation has shown that differences with more factors have more complex patterns.

Problem #10

10. Create a general method to predict the exact pattern of greatest common factors for any difference (without actually calculating any of the GCFs). Illustrate your method on pairs that have a difference of 12.

Questions and Conversations for #10

» *How long does it take for each pattern to repeat?* If the difference is n, the pattern repeats after every n numbers.
» *What numbers belong to the pattern?* The numbers in the pattern are the factors of the difference. (See Problem #6.)

» *What is the smallest number in the pattern? How often is it a common factor?*
The smallest number in the pattern is always 1. It's a common factor of every pair. We might begin to show the pattern like this:

$$1, 1, 1, 1, 1, 1, 1, \ldots$$

» *What is the next larger number in the pattern? How often is it a common factor?*
This depends upon the difference. For example, if it is even, the next larger number in the pattern will be 2 and it will be a common factor of every second pair. We might replace every second 1 with the larger common factor 2:

$$1, 2, 1, 2, 1, 2, 1, \ldots$$

Solution for #10

This can be a challenging problem. See the Questions and Conversations for ideas to help students get started.

If the difference is n, the pattern repeats every n numbers in the list. The pattern for a difference of 12 repeats this block of 12 numbers:

$$1, 2, 3, 4, 1, 6, 1, 4, 3, 2, 1, 12, \ldots$$

Here is one strategy for predicting the pattern without calculating the GCFs:
The list will contain only factors of 12. Because each of the 12 pairs has a common factor of 1, we begin with:

$$1, 1, 1, 1, 1, 1, 1, 1, 1, 1, 1, 1$$

Every second pair has a common factor of 2:

$$1, 2, 1, 2, 1, 2, 1, 2, 1, 2, 1, 2$$

Every third pair has a common factor of 3:

$$1, 2, 3, 2, 1, 3, 1, 2, 3, 2, 1, 3$$

Every fourth pair has a common factor of 4:

$$1, 2, 3, 4, 1, 3, 1, 4, 3, 2, 1, 4$$

Every sixth pair has a common factor of 6:

$$1, 2, 3, 4, 1, 6, 1, 4, 3, 2, 1, 6$$

And finally, every twelfth pair has a common factor of 12:
$$1, 2, 3, 4, 1, 6, 1, 4, 3, 2, 1, 12$$

Notice how we replaced smaller common factors by larger ones at each step to produce the greatest common factor? Another approach is to begin with the largest factor and then work our way down. This eliminates the need for erasing and replacing the smaller factors.

STAGE 3

Teacher's Note for #11. Students' experience with the connections between differences and greatest common factors should help them structure their investigation of least common multiples. If they don't see any patterns in their data, suggest that they compare the LCM to the product of the numbers in each pair.

Problem #11

11. Search for a connection between differences and least common multiples. Describe any discoveries that you make.

Solution for #11

We can create organized lists, just like before.

Differences of 1		Differences of 2		Differences of 3	
Number Pair	LCM	Number Pair	LCM	Number Pair	LCM
1, 2	2	1, 3	3	1, 4	4
2, 3	6	2, 4	4	2, 5	10
3, 4	12	3, 5	15	3, 6	6
4, 5	20	4, 6	12	4, 7	28
5, 6	30	5, 7	35	5, 8	40
6, 7	42	6, 8	24	6, 9	18

When the numbers differ by 1, the LCM is the product of the numbers.

When the numbers differ by 2, the LCM is the product except every second time.

When the numbers differ by 3, the LCM is the product except every third time.

Some students might observe that when the LCM is not the product, it is instead the product divided by the difference! However, if you test differences that aren't prime numbers, the pattern is a little more complicated. The number that the product is divided by each time follows these familiar patterns:

$$1, 2, 1, 2, 1, 2, 1, 2, 1 \text{ for differences of 2}$$

$$1, 1, 3, 1, 1, 3, 1, 1, 3 \text{ for differences of 3}$$

$$1, 2, 1, 4, 1, 2, 1, 4, 1 \text{ for differences of 4}$$

It looks like the difference is affecting the LCM via the GCF! (See Exploration 9: The GCF-LCM Connection.)

WRAP UP

Share Strategies

Ask students to share and compare their strategies and observations.

Summarize

Answer remaining questions that students have. Then summarize these key points:

» Greatest common factors are always factors of the difference of two numbers.
» When pairs are organized into lists, we can see patterns in the greatest common factors. We can also understand what causes these patterns by analyzing factors of equally spaced numbers on the number line. (See Questions and Conversations for #6.)

Practice simplifying more fractions using the difference-GCF connection.

Further Exploration

Try this experiment:
1. Pick any two natural numbers and find their positive difference (the larger number minus the smaller).
2. Take the two smaller numbers from the previous calculation and find their positive difference. (If the calculation was $a - b = c$, then the two smaller numbers will be b and c.)
3. Continue this process until it ends at 0—or until you can see that it will never end.

Think about these questions:
» Does the process always end? If so, why? If not, what will make it go on forever?
» What does each pair of numbers that you subtracted have in common? Why?
» What do you notice about the final two numbers that you subtracted (besides the fact that they are equal)? Again, why?
» How can you make the process go on for a long time?
» How does all of this relate to the concepts in this exploration?
» What is the *Euclidean Algorithm*? How does it relate to this experiment and to this whole activity? Do some research!

Exploration 7

A Measurement Dilemma

INTRODUCTION

Prior Knowledge

>> Know the definition of *greatest common factor*.

>> Know that the product $a \cdot b$ can be represented as ab.

>> Understand that *solving an algebraic equation* means finding the values of the variables that make it true (Stages 2 and 3). (*Note.* Students do not have to know procedures for solving linear equations.)

Learning Goals

>> Develop strategies for solving problems involving linear combinations, $ax - by$.

>> Gather and organize data. Use it to make and test mathematical conjectures.

>> Explore connections between linear combinations and greatest common factors.

>> Use algebraic expressions to model a physical situation.

>> Develop strategies and models to find whole number solutions of equations.

>> Analyze patterns in solutions to algebraic equations.

>> Communicate complex mathematical ideas clearly.

>> Persist in solving challenging problems.

Launching the Exploration

Motivation and purpose. To students: You are about to tackle a well-known and popular mathematical puzzle. However, you won't stop once you've found a solution or two. Instead, you'll change the numbers and keep trying! The goal is to discover ways to solve the problem quickly no matter what numbers you are given. As you work, you will make some surprising connections to familiar mathematics as well as to some new ideas from algebra.

Understanding the problem. Read the problem carefully. The only actions allowed are those stated. For example, you may not place containers side by side to compare water levels, write on the containers, estimate a fraction of a container, etc.

Read through the entire activity to help students see the big picture. Use the first Questions and Conversations to guide an introductory discussion. In Stage 1, students explore the problem with different pairs of containers. Then they use what they learn to

DOI: 10.4324/9781003232742-9

make a general conjecture about the situation. In the remainder of the exploration, they apply their new knowledge to develop strategies and models for finding whole number solutions to algebraic equations.

Have students start the activity with your help. Check that they understand the constraints of the problem and are prepared to work on their own. Expect that some students may finish Problem #1 quite quickly, while others will take much longer.

STUDENT HANDOUT

You have a 5-cup container and a 7-cup container. The containers are unmarked and opaque. You are allowed to do three things:
  » Fill either container to the top from the faucet.
  » Transfer as much water as possible from one container to another.
  » Empty either container completely.

Stage 1

1. Find a way to measure exactly 6 cups in one of the containers. After you find one solution, look for another! Describe any patterns that you see in your solutions. (*Note.* If one of your containers holds extra water at the end, always empty it out.)

2. Suppose you have a third container (as large as you'd like) so that it's possible to store amounts larger than 7 cups. What amounts can be measured? Explain your thinking.

3. For each solution in Problem #1, how many times did you fill and empty each container? How can you see from this that the final result must be 6 cups?

4. Continue the investigation for other pairs of containers, trying to discover all possible amounts that can be measured. Describe your investigations and explain your thinking. (Suggested container sizes: 6 cups and 10 cups; 9 cups and 21 cups)

5. Analyze the results from your investigations and make a conjecture. What amounts can you measure with an *x*-cup and a *y*-cup container? Explain your thinking. Test your conjecture by trying it on at least one more pair of containers.

6. Explain how to use your answer to Problem #3 to see a solution to the equation $5x - 7y = 6$.

7. Suppose you get 6 cups of water (with the 5- and 7-cup containers) by measuring 1 cup, storing it in a third container, and then repeating this process 6 times. Use this to find a new solution to $5x - 7y = 6$. Explain your thinking.

8. Compare your two solutions. Show how to find a third solution between the two.

9. Find a pattern in your three solutions. Predict three larger solutions. Explain how you did this and why this process works. How many more solutions do you think there are?

10. Continue your pattern in the opposite direction to find two solutions that involve negative numbers. Then use these negative number solutions to find two non-negative solutions to $7y - 5x = 6$. Compare one of the solutions to one of your answers to Problem #1. What do you notice?

11. Starting at the top, imagine skip-counting your way around this circle, jumping 5 dots each time. Explain how you could use the circle to solve Problem #1. *Note.* Include the number 1 as your first jump. (You should land on 5 the first time.)

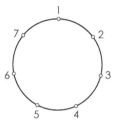

12. Show how to use a circle like this to solve the equation $13x - 20y = 17$. Then explain how to use what you learned in Stage 2 to find three more solutions.

Advanced Common Core Math Explorations: Factors & Multiples © Taylor & Francis.

TEACHER'S GUIDE

You have a 5-cup container and a 7-cup container. The containers are unmarked and opaque. You are allowed to do three things:

» Fill either container to the top from the faucet.
» Transfer as much water as possible from one container to another.
» Empty either container completely.

Questions and Conversations for the Introduction

This section contains ideas for conversations, mainly in the form of questions that students may ask or that you may pose to them. Be sure to allow students to do most of the thinking and talking!

» *What does "opaque" mean?* It is the opposite of transparent. You can't see through the sides of the containers. (This prevents you from placing the containers side by side and comparing water levels.)

» *Can you use a third container to store water?* See if you can solve the problem without a third container for now. This makes it a little more challenging and interesting.

» *Can it help to act it out?* It might! You may not have any 5-cup and 7-cup containers handy, but you can pretend!

» *Why is it important to keep track of what you've tried?* There are many reasons. It prevents wasting time by repeating failed attempts. It gives you a way to backtrack and correct errors. It can help you find patterns.

» *What's the best way to keep track of what you've tried?* Lists or tables might help. Some people like to create abbreviations for the actions such as "F5" for filling the 5-cup container, or to draw diagrams illustrating the actions, showing how many cups are in the containers at each step.

STAGE 1

Problem #1

1. Find a way to measure exactly 6 cups in one of the containers. After you find one solution, look for another! Describe any patterns that you see in your solutions. (*Note.* If one of your containers holds extra water at the end, always empty it out.)

Questions and Conversations for #1

» *How many different action choices do you have in each step?* There are really only six choices. Fill the 5, empty the 5, fill the 7, empty the 7, transfer from 5 to 7, or transfer from 7 to 5.

» *Are there any key strategies for deciding the best next move?* This is what the problem is all about! Consider every choice. Try to avoid undoing actions that you've already taken. You might be surprised at how easy it can be to decide.

» *What if you get every possible amount except 6 cups?* Don't stop too soon! You may be closer to getting 6 cups than you think. People often give up and start over right before they would reach a solution.

» *What's the best way to communicate the solution?* The methods that you use to keep track of your own work may often help communicate it to others as well. Tables, lists, and diagrams can all be effective. Paragraphs of purely verbal explanations may be harder to follow.

» *What counts as a different solution?* If a solution contains many of the same actions, but has extra steps that just undo each other, you probably shouldn't count it as a new solution. Consider starting by filling a different container than you did the first time.

Solution for #1

We'll show the process using tables: F stands for "fill," E stands for "empty," and the arrows represent transfers. The numbers in the bottom two rows of each table show the amounts in the 5- and 7-cup containers at each step.

Action	F5	5→7	F5	5→7	E7	5→7	F5	5→7	E7	5→7	F5	5→7
5-cup	5	0	5	3	3	0	5	1	1	0	5	0
7-cup	0	5	5	7	0	3	3	7	0	1	1	6

Action	F7	7→5	E5	7→5	F7	7→5	E5	7→5	F7	7→5	E5
5-cup	0	5	0	2	2	5	0	4	4	5	0
7-cup	7	2	2	0	7	4	4	0	7	6	6

Patterns: You always alternate filling or emptying with transferring. The first solution contains only three actions: F5, 5→7, and E7. The second solution contains only the other three actions: F7, 7→5, and E5.

Problem #2

2. Suppose you have a third container (as large as you'd like) so that it's possible to store amounts larger than 7 cups. What amounts can be measured? Explain your thinking.

Questions and Conversations for #2

» *What is the smallest amount you can measure with a 5-cup and 7-cup container? How do you know?* You can measure 1 cup. You actually got this while solving the original problem!

» *How can you find all measurable amounts once you know the smallest amount?* Once you have 1 cup, you can store it in the third container. How can you proceed from there?

Teacher's Note. Students who look ahead to Problem #7 may see that it practically contains the answer to this question. It's fine if they notice this, but you don't have to point it out to them!

Solution for #2

The first solution shows that you can make 1 cup. By storing this in the third container and repeating the process as many times as necessary, you can measure any whole number amount (although this will not generally be the most efficient method).

Problem #3

3. For each solution in Problem #1, how many times did you fill and empty each container? How can you see from this that the final result must be 6 cups?

Questions and Conversations for #3

» *How does transferring water between containers affect the amount you have?* It doesn't!

Teacher's Note. Make sure students have emptied any extra water in the second container as mentioned in the note for Problem #1.

Solution for #3

In the first solution, you filled the 5-cup container 4 times and emptied the 7-cup container twice, leaving a total of $5 \cdot 4 - 7 \cdot 2 = 20 - 14 = 6$ cups. (Transferring between containers does not change the total.)

In the second solution, you filled the 7-cup container 3 times, and emptied the 5-cup container 3 times, leaving a total of $7 \cdot 3 - 5 \cdot 3 = 21 - 15 = 6$ cups.

Problem #4

4. Continue the investigation for other pairs of containers, trying to discover all possible amounts that can be measured. Describe your investigations and explain your thinking. (Suggested container sizes: 6 cups and 10 cups; 9 cups and 21 cups)

Questions and Conversations for #4

» *Can you still find two solutions for each pair of measuring cups?* Yes. Can you see why?

» *Can you still measure 1 cup with the suggested container sizes?* No. Finding the smallest measurable amount is the key!

Solution for #4

Start by finding the smallest whole number of cups you can measure. These tables show two ways of measuring 2 cups.

Action	F6	6→10	F6	6→10	E10
6-cup	6	0	6	2	2
10-cup	0	6	6	10	0

Action	F10	10→6	E6	10→6	F10	10→6	E6	10→6	E6
6-cup	0	6	0	4	4	6	0	6	0
10-cup	10	4	4	0	10	8	8	2	2

You can't measure an amount smaller than this because 6 and 10 are both even. Sums and differences of even numbers are always even.

If you repeat the process with 9- and 21-cup containers, you will be able to measure 3 cups. Because 9 and 12 are both multiples of 3, and sums and differences of multiples of 3 are still multiples of 3, this is the smallest amount you can measure.

As before, if you make the smallest amount possible, store it in the third container, and repeat the process, you can make any multiple of this. Therefore, you can measure any even number of cups with the 6- and 10-cup containers and any multiple of 3 with the 9- and 21-cup containers.

Problem #5

5. Analyze the results from your investigations and make a conjecture. What amounts can you measure with an x-cup and a y-cup container? Explain your thinking. Test your conjecture by trying it on at least one more pair of containers.

Questions and Conversations for #5

» *How can you find amounts that will work for x-cup and y-cup containers when you don't know what the values of x and y are?* Find a rule that will work for any natural numbers x and y. You can write your answer in words or with a formula.

> **Teacher's Note.** Tell students not to jump too quickly to conclusions. If they're having trouble, suggest that they keep experimenting with containers of other sizes.

Solution for #5

The amounts that you can measure are the multiples of the greatest common factor of the containers' sizes.

STAGE 2

Problem #6

6. Explain how to use your answer to Problem #3 to see a solution to the equation $5x - 7y = 6$.

Questions and Conversations for #6

» *How does the equation $5x - 7y = 6$ connect to the problem situation? What could the numbers 5, 7, and 6 stand for?* The numbers could stand for the container sizes and the amount that you're measuring. *What about x and y?*

Solution for #6

In Question 3, you saw that $5 \cdot 4 - 7 \cdot 2 = 6$. This means that $x = 4, y = 2$ is a solution of the equation $5x - 7y = 6$.

Problem #7

7. Suppose you get 6 cups of water (with the 5- and 7-cup containers) by measuring 1 cup, storing it in a third container, and then repeating this process 6 times. Use this to find a new solution to $5x - 7y = 6$. Explain your thinking.

Questions and Conversations for #7

» *How many "empties" and "fills" did it take to make 1 cup? What happens to these numbers when you repeat the process six times?* The answer to the first question is somewhere in your solution to Problem #1!

Solution for #7

Your work in Problem #1 shows that you can measure 1 cup by filling the 5-cup container 3 times and emptying the 7-cup container 2 times. To make 6 cups, you will fill and empty each container 6 times as many times as you did before. Therefore, another solution is $x = 18, y = 12$.

Problem #8

8. Compare your two solutions. Show how to find a third solution between the two.

Solution for #8

The solutions for x were 4 and 18, and for y were 2 and 12. It turns out that you can choose the numbers exactly in the middle of each pair: $x = 11, y = 7$:

$$5 \cdot 11 - 7 \cdot 7 = 55 - 49 = 6$$

Problem #9

9. Find a pattern in your three solutions. Predict three larger solutions. Explain how you did this and why this process works. How many more solutions do you think there are?

Questions and Conversations for #9

» *What kind of pattern did you create when you found a solution exactly in the middle of the first two?* The pattern might be easier to see if you organize the solutions in a list or table.

Solution for #9

This table shows that each value of x increases by 7, while y increases by 5.

x	y
4	2
11	7
18	12

Based on this pattern, you can predict the three larger solutions:

$x = 25 \; y = 17$ $x = 32 \; y = 22$ $x = 39 \; y = 27$

You can verify these by doing the calculations. For example:

$$5 \cdot 25 - 7 \cdot 17 = 125 - 119 = 6$$

This works because increasing x by 7 adds 7 groups of 5, while increasing y by 5 subtracts 5 groups of 7, leaving you with the same solution. There will be infinitely many solutions because you can continue this pattern forever.

Problem #10

10. Continue your pattern in the opposite direction to find two solutions that involve negative numbers. Then use these negative number solutions to find two non-negative solutions to $7y - 5x = 6$. Compare one of the solutions to one of your answers to Problem #1. What do you notice?

Questions and Conversations for #10

» *What is the relationship between the equations $5x - 7y = 6$ and $7y - 5x = 6$?* The order of subtraction is reversed on one side.

Teacher's Note. Even if students aren't yet familiar with negative number computation, they can probably make reasonable conjectures about how to add, subtract, and multiply with the negative solutions—especially because they know what to expect for an answer!

Solution for #10

If you begin at $x = 4$, $y = 2$ and extend the pattern backward, subtracting 7 from x and 5 from y, you get

$$x = -3 \quad y = -3 \qquad x = -10 \quad y = -8$$

You can use these to get solutions to the equation $7y - 5x = 6$ by taking the opposites of these x and y values.

$$7 \cdot 3 - 5 \cdot 3 = 6 \text{ and } 7 \cdot 8 - 5 \cdot 10 = 6$$

The first of these solutions matches with the answer to Problem #1, where you filled the 7-cup container 3 times and emptied the 5-cup container 3 times.

STAGE 3

Problem #11

11. Starting at the top, imagine skip-counting your way around this circle, jumping 5 dots each time. Explain how you could use the circle to solve Problem #1. *Note.* Include the number 1 as your first jump. (You should land on 5 the first time.)

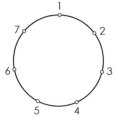

Questions and Conversations for #11

» *What data might you record as you move around the circle?* How many times did you "skip"? Which numbers did you land on? How many times did you land on or cross the top of the circle? In what order did these things happen?

Solution for #11

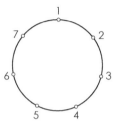

Start at the top and count clockwise by fives until you land on 6, the desired number of cups. Each time you jump by five, it stands for filling the 5-cup container. Each time you pass or land on the number 1, it represents emptying the 7-cup container.

If you jump four times, passing or landing on the number 1 twice along the way, you will stop at 5, 3, 1, and then 6. This represents filling the 5-cup container four times and emptying the 7-cup container twice, which agrees with the original solution.

To get the exact series of steps as they appear in your table, pay attention to the order in which things happen as you travel around the circle and insert a transfer (5→7) between each step.

Note. If you use a circle to solve the problem by filling the 7-cup container first, you will have to think a little more carefully, because your circle will have only 5 numbers on it. You will actually have 6 cups when you land on the number 1. Can you see why?

Problem #12

12. Show how to use a circle like this to solve the equation $13x - 20y = 17$. Then explain how to use what you learned in Stage 2 to find three more solutions.

Questions and Conversations for #12

» *How can you redraw the circle and adjust the skips to fit this equation?* Pay attention to the containers' sizes.

Solution for #12

Make a circle with the numbers 1 through 20 and skip-count by 13 until you hit 17. In the process, you land on 13, 6, 19, 12, 5, 18, 11, 4, and 17. (Notice the patterns!)

In the process, you carry out 9 jumps of 13, crossing or landing on the number 1 at the top of the circle 5 times. This represents the solution $x = 9, y = 5$, which can be verified:

$$13 \cdot 9 - 20 \cdot 5 = 117 - 100 = 17$$

You can find three more solutions by adding 20 to x, and 13 to y each time.

$x = 29\ y = 18$ $\qquad\qquad x = 49\ y = 31$ $\qquad\qquad x = 69\ y = 44$

WRAP UP

Share Strategies

Allow students to share and compare their strategies and observations.

Summarize

Answer any remaining questions that students have. Discuss a few points related to questions that the students worked on:

» Solving an algebraic equation means to find values that make the equation true. There are many methods for doing this depending on the type of equation you have and the type of solution you're looking for.

» There are connections between greatest common factors, the process of adding and subtracting multiples, and "circular addition." These ideas can also be used to find whole number solutions to certain types of algebraic equations.

Further Exploration

Have students think of new questions to ask or ways to continue or extend this exploration. Here are some possibilities:

» Experiment with solving equations like those in this activity, using addition instead of subtraction: $ax + by = c$.

» Continue exploring the idea of "skip counting" around a circle. Try a variety of number pairs. Find a connection to *least common multiples*.

» Explore connections to the drawing toy, *Spirograph*®. How do you think the makers chose the number of teeth on the rings and wheels? How can you predict the number of "tips" a drawing will have and how many times you go around a circle to complete a design?

» If you are interested in music, do some research on the circle of fifths. How does it relate to the mathematical concepts in this exploration? Why do you cycle through all 12 keys before coming back to where you started?

» Do some research on Diophantine equations.

Exploration 8

Paper Pool

INTRODUCTION

Materials

» Graph paper

Prior Knowledge

» Calculate *greatest common factors* (GCF) and *least common multiples* (LCM).
» Use *ratios* to compare lengths.
» Compute fluently with fractions and decimals (Stage 3).

Learning Goals

» Use visual models to deepen understanding of GCFs and LCMs.
» Organize data to analyze and extend patterns.
» Make connections between GCFs, LCMs, and ratios.
» Use ratios to analyze size changes and solve scaling problems.
» Enhance spatial visualization skills.
» Communicate complex mathematical ideas clearly.
» Persist in solving challenging problems.

Launching the Exploration

Motivation and purpose. To students: In this activity, you will investigate paths of a ball on a pool table. When a ball comes in at the side of the table from a certain angle, it bounces off at the same angle. When the angle is 45°, you can use graph paper to help you sketch the paths. Many surprising and intricate patterns arise, and your job is to explore them!

Understanding the problem. Use the example of the 10 by 8 grid on the student handout to demonstrate how the ball moves in Paper Pool. Introduce the terminology: *dimensions, path length, hit,* and *gap length.*

Show students that a *diagonal unit* (used to measure path length) is longer than a *unit.* Point out that hits include the points where the ball starts and stops. Also, the gap length between the hits at B and C in the example is 4 units, even though you go around a corner.

 DOI: 10.4324/9781003232742-10

Assist students as needed when they begin the activity. Check that they are drawing the paths correctly and that they can identify path lengths, gap lengths, and hits.

Teacher's Notes. Don't tell students that the exploration is about greatest common factors, least common multiples, and ratios. Allow them to discover this for themselves.

This exploration was inspired by an enrichment activity in *Everyday Mathematics* (Bell et al., 2007).

STUDENT HANDOUT

Rules for Paper Pool: Start in the lower left corner of the table. Travel at a 45° angle to the sides. When you hit a side, bounce off at a 45° angle. When you reach a corner, the ball falls into the pocket and you stop. Here's an example on a 10 by 8 pool table (grid).

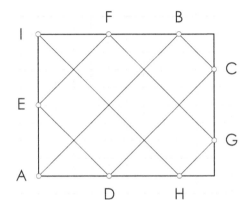

Terminology:

- » *Dimensions*: the length and width of a grid
- » *Path length*: the number of diagonal units traveled by the ball
- » *Hit*: a point where the ball touches a side or a corner, including the starting and ending points
- » *Gap length*: the number of units between nearest hits

Notice that a diagonal unit (from corner to corner) is longer than a unit (the length of a side):

To find the path length, count the number of diagonal units from the beginning of the path until the end. To find the gap length, count the number of units along the edge of the table between any two neighboring hits (even if they are around a corner from each other). The path above has 9 hits, gap lengths of 4 units, and a path length of 40 diagonal units.

1. Use graph paper to draw paths on Paper Pool tables with these dimensions.

   4 by 3   5 by 2   8 by 3   9 by 6   10 by 6   10 by 2   9 by 3

2. Create a mathematical table to collect and organize data for each path.

3. Discover a rule to find the path length (*P*) on an *a* by *b* Paper Pool table. Explain your thinking.

Now you will analyze other features of your data, including hits and gap lengths.
4. Make drawings for Paper Pool tables with these dimensions. Create a mathematical table similar to the one you made in Problem #2 for these paths.

   12 by 4   5 by 1   20 by 8   3 by 2   8 by 6   5 by 3   15 by 10

5. Sort all 14 of your drawings into categories of paths that have the same shape. Create a separate table for each category. Organize the rows in each table according to the sizes of the paths. Fill in rows for any missing paths, so that each table contains at least the three or four smallest paths its category.

6. Describe any patterns you see in the values of *a* and *b* within each category.

7. Let *N* represent the size of a path within a category. For example, *N* = 1 stands for the shortest path, *N* = 2 represents the second shortest path, etc. Explain how to use the dimensions, *a* and *b*, to predict the value of *N* for any path. What more specifically does *N* tell you about the length of a path?

8. Find and describe at least two or three interesting patterns and relationships involving hits and gap lengths.

9. Choose some patterns and relationships from Problem #8 and explain what causes them or why they make sense.

10. Find a formula for the shortest path length (P) in a category using only the variables a, b, and N. Explain your thinking. Then, show how to adjust your formula so that it works for any path.

11. N and P are connected to two mathematical concepts that you have seen before. In your formula, replace N and P with names or abbreviations of these concepts. Then test your formula on at least two pairs of values for a and b.

Stage 3

12. Analyze a Paper Pool table with dimensions of 3.5 units and $4\frac{2}{3}$ units. Show how to find the path length (P), hits (H), and gap lengths (G). Sketch the path.

13. Design a Paper Pool table whose path has gap lengths of 3.4 units and a path length of 25.5 units. What are the dimensions? How many hits will there be? Sketch the path. Explain your thinking process.

TEACHER'S GUIDE

Rules for Paper Pool: Start in the lower left corner of the table. Travel at a 45° angle to the sides. When you hit a side, bounce off at a 45° angle. When you reach a corner, the ball falls into the pocket and you stop. Here's an example on a 10 by 8 pool table (grid).

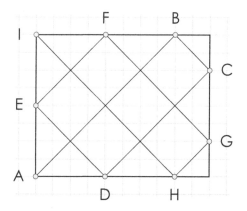

Terminology:

- » *Dimensions*: the length and width of a grid
- » *Path length*: the number of diagonal units traveled by the ball
- » *Hit*: a point where the ball touches a side or a corner, including the starting and ending points
- » *Gap length*: the number of units between nearest hits

Notice that a diagonal unit (from corner to corner) is longer than a unit (the length of a side):

To find the path length, count the number of diagonal units from the beginning of the path until the end. To find the gap length, count the number of units along the edge of the table between any two neighboring hits (even if they are around a corner from each other). The path above has 9 hits, gap lengths of 4 units, and a path length of 40 diagonal units.

STAGE 1

Problem #1

1. Use graph paper to draw paths on Paper Pool tables with these dimensions.

    4 by 3   5 by 2   8 by 3   9 by 6   10 by 6   10 by 2   9 by 3

Questions and Conversations for #1

This section contains ideas for conversations, mainly in the form of questions that students may ask or that you may pose to them. Be sure to allow students to do most of the thinking and talking!

» *How can you check that your paths are exactly at 45° angles to the sides?* Paths that run at 45° angles to the sides go from one corner to the opposite corner of each grid square.

Solution for #1

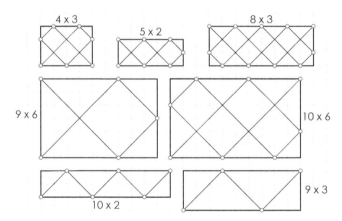

Problem #2

2. Create a mathematical table to collect and organize data for each path.

Questions and Conversations for #2

» *What should be the columns in the table?* You should probably have columns for the dimensions *a* and *b*, the path lengths, the hits, and the gaps. You may also think of additional data to include.

Solution for #2

The following data should appear in students' tables. They may choose to include other data as well. For example, they may have columns for perimeter, area, number of gaps, symmetries of the paths, the corners that the balls land in, or the numbers of intersection points of the paths.

a	b	hits	gap length	path length
4	3	7	2	12
5	2	7	2	10
8	3	11	2	24
9	6	5	6	18
10	6	8	4	30
10	2	6	4	10
9	3	4	6	9

Problem #3

3. Discover a rule to find the path length (*P*) on an *a* by *b* Paper Pool table. Explain your thinking.

Questions and Conversations for #3

» *When does the path length equal the larger of a and b?* This happens when one dimension is a multiple of the other.

» *When does the path length equal the product of a and b?* This happens when *a* and *b* have no common factors except 1.

» *What would happen if you extended the grid so the ball kept traveling to the right instead of bouncing back?* It would look like this example of an "extended grid" for the 10 by 8 table:

Teacher's Note. Students might need to draw more paths in order to reach these conclusions. Then they can search for a single concept or calculation that fits both conclusions. If they still have trouble, try asking questions like the ones listed for Questions and Conversations for #3.

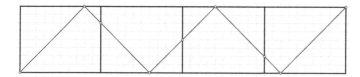

> » *How does the path length on this "extended grid" compare to the original path length?* They are the same—40 diagonal units in this case. Can you see why?
> » *This path is made of 5 segments. How can you predict from the dimensions of the table how far each segment moves to the right?* Each segment moves to the right an amount equal to the width (*b*) of the table—8 units in this case. Why does this happen?
> » *How can you predict how many copies of the original Paper Pool table you will need to use in the extended grid?* You will continue until the total length of the extended grid is a multiple of *b* (the amount that each segment of the path moves to the right).

Solution for #3

The path length is equal to the least common multiple of *a* and *b*.

You can see why by drawing an extended grid so that the ball keeps moving to the right. This picture shows an example of an extended grid using a 10 by 8 table. The path is broken into 5 segments, each of which moves 8 units to the right (the same as the width of the table, *b*). Each copy of the table has a length

> **Teacher's Note for #3.** Some students may discover a formula that involves the greatest common factor of *a* and *b* instead. If this happens, use the extended grid to guide them toward the least common multiple relationship, and let them know that their result will appear later in Problems #10 and #11!

(*a*) of 10 units. The ball hits a corner at 40 units, the first time that a multiple of 8 (*b*) is equal to a multiple of 10 (*a*). This is the least common multiple of 8 and 10.

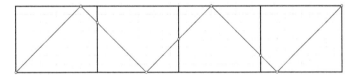

Students may notice other things: When one dimension is a multiple of the other, the path length is the larger dimension. When the dimensions have no common factor other than 1, the path length is the product of the dimensions.

Problem #4

Now you will analyze other features of your data, including hits and gap lengths.

4. Make drawings for Paper Pool tables with these dimensions. Create a mathematical table similar to the one you made in Problem #2 for these paths.

12 by 4   5 by 1   20 by 8   3 by 2   8 by 6   5 by 3   15 by 10

Solution for #4

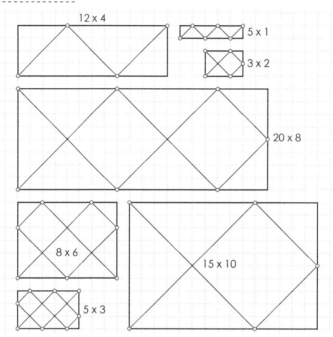

a	*b*	hits	gap length	path length
12	4	4	8	12
5	1	6	2	5
20	8	7	8	40
3	2	5	2	6
8	6	7	4	24
5	3	8	2	15
15	10	5	10	30

Problem #5

5. Sort all 14 of your drawings into categories of paths that have the same shape. Create a separate table for each category. Organize the rows in each table according to the sizes of the paths. Fill in rows for any missing paths, so that each table contains at least the three or four smallest paths in its category.

Questions and Conversations for #5

» *What does it mean for paths to have the "same shape"?* Different paths with the same shape should look like enlargements or reductions of each other. (These are called *similar* figures.)

» *What does the phrase "missing paths" mean?* This is easier to see after you have ordered the paths in each category by size. You should be able to imagine paths with sizes intermediate between the ones you have already drawn.

> **Teacher's Note.** You could introduce or review the term *similar* at this point.

» *How do you know what data to enter for these paths if you haven't drawn them?* Consider using patterns in your tables to guide you. And, of course, it's always a good idea to draw them to verify your predictions!

Solution for #5

There will be seven categories. The rows in gray were added to help complete patterns and make them more visible.

a	b	hits	gap length	path length
4	3	7	2	12
8	6	7	4	24
12	9	7	6	36
16	12	7	8	48

a	b	hits	gap length	path length
3	1	4	2	3
6	2	4	4	6
9	3	4	6	9
12	4	4	8	12

a	b	hits	gap length	path length
5	2	7	2	10
10	4	7	4	20
15	6	7	6	30
20	8	7	8	40

a	b	hits	gap length	path length
5	1	6	2	5
10	2	6	4	10
15	3	6	6	15
20	4	6	8	20

a	b	hits	gap length	path length
3	2	5	2	6
6	4	5	4	12
9	6	5	6	18
12	8	5	8	24
15	10	5	10	30

a	b	hits	gap length	path length
5	3	8	2	15
10	6	8	4	30
15	9	8	6	45
20	12	8	8	60
25	15	8	10	75

a	b	hits	gap length	path length
8	3	11	2	24
16	6	11	4	48
24	9	11	6	72

Problem #6

6. Describe any patterns you see in the values of a and b within each category.

Questions and Conversations for #6

» *What can you say about the values of a and* b *for the shortest path in each category?* Not surprisingly, the smallest values of a and b belong to the shortest paths.

» *How does each value of* b *compare to its matching value of* a *within a category?* Try comparing the values of a and b by using different operations. The patterns may be easier to see in some categories than others.

» *How can you change one pair into another within the same category?* Experiment to find a process that always works. The patterns may be easier to see between some pairs than others.

Solution for #6

Sample student responses:

» The ratio of a to b (a divided by b) is the same for each path in the category.

» Starting with the smallest path in a category, you can get the other paths by multiplying its dimensions by 2, 3, 4, etc.

» If you use the values of a and b to form fractions, then the fractions in each category will be equivalent.

» Smaller paths have smaller values of a and b within a category. The dimensions of the smallest path in a category always have a greatest common factor of 1, and their product is equal to the path length.

Problem #7

7. Let N represent the size of a path within a category. For example, $N = 1$ stands for the shortest path, $N = 2$ represents the second shortest path, etc. Explain how to use the dimensions, a and b, to predict the value of N for any path. What more specifically does N tell you about the length of a path?

Questions and Conversations for #7

» *Can you write the answer as a formula?* Possibly—though you might have more success thinking of it as a process.

> » *What does it mean to be "more specific" about the connection between N and the path lengths?* N tells you more than just the rank order of the paths.

Solution for #7

The value of *N* is equal to the greatest common factor of *a* and *b*, (GCF(*a*,*b*)).

Each path length is *N* times as long as the shortest path in the category. For example, if *a* = 8 and *b* = 6, then *N* = GCF(8, 6) = 2 and the path length is 24, which is 2 times greater than 12, the path length for the 4 by 3 table.

Problem #8

8. Find and describe at least two or three interesting patterns and relationships involving hits and gap lengths.

Solution for #8

Sample student responses:

> » The gap length for the smallest path in each category is always 2.
> » The gap length is $2 \cdot N$, or $2 \cdot GCF(a, b)$ for every path.
> » The number of hits in a category is the same for every path in a category. It is equal to the value of $a + b$ for the smallest path.
> » The number of hits is the same as the number of gaps for all paths.
> » The perimeter of the Paper Pool table is equal to the gap length multiplied by the number of hits (or gaps).
> » Increasing the dimensions by some factor increases the gap length (and the path length) by that same factor. For example, if you double the length and width of a table, you also double the gap length and the path length.

Teacher's Note for #8. Here are some questions to focus students' thinking if they are having trouble knowing what to look for.
- Do you see patterns in the hits or gap lengths within a category?
- How can you use the dimensions of the grids to predict hits and gap lengths?
- How does the number of gaps relate to the number of hits?
- Can it help to focus on the perimeter of the grid?
- How do hits and gap lengths relate to other data such as path lengths?

STAGE 2

Problem #9

9. Choose some patterns and relationships from Problem #8 and explain what causes them or why they make sense.

Questions and Conversations for #9

» *Can the "extended grids" help? (See Questions and Conversations for #3.)* They might be especially helpful for analyzing the number of hits.

Solution for #9

Sample student response. Why is the number of hits equal to the value of $a + b$ for the smallest table in the category? You can use an extended grid for the smallest table to understand why. Ignore the start and end points for now. The smallest path is always made of a segments (can you see why?), so there are $a - 1$ hits between them at the top and bottom of the table.

The number of hits on the left and right sides of the table is $b - 1$ because there are b copies of the table, and $b - 1$ is the number of times the ball crosses the border between them.

If you add everything together, you get
top/bottom hits + left/right hits + start/end hits =
$(a - 1) + (b - 1) + 2 = a + b$ hits

for the shortest path in the category. The number of hits for the longer paths is the same because these paths are just enlargements of the smallest one.

Sample student response. Why is the number of gaps the same as the number of hits? Think about why the number of vertices of a polygon is equal to its number of sides. The reason for this is essentially the same.

Sample student response. Why is the gap length $G = 2 \cdot GCF(a,b)$? The gap length for the smallest grid in each category is 2, because it is the perimeter divided by the number of gaps. The perimeter is $2 \cdot (a + b)$ and the number of gaps is $a + b$. This gives a gap length of $2 \cdot (a + b) \div (a + b)$, which equals 2. You multiply this by N (the GCF of a and b) for the other grids in the category to give a formula of $G = 2 \cdot GCF(a,b)$.

Problem #10

10. Find a formula for the shortest path length (P) in a category using only the variables a, b, and N. Explain your thinking. Then, show how to adjust your formula so that it works for any path.

Questions and Conversations for #10

» *How can you adjust the values of* a *and* b *to get the dimensions of the table with the shortest path in a category?* Remember that the dimensions are N times greater than those for the shortest path in the category.

Solution for #10

Divide both a and b by N to get the dimensions of the table with the shortest path in the category. Then multiply these to get the length of the shortest path.

$$(a \div N) \cdot (b \div N)$$

Multiply this answer by N to get the length of the original path.

$$P = (a \div N) \cdot (b \div N) \cdot N$$

Some students may discover that they can write this in simpler ways. For example:

$$P = a \div N \cdot b \text{ or } P = a \cdot b \div N$$

Problem #11

11. N and P are connected to two mathematical concepts that you have seen before. In your formula, replace N and P with names or abbreviations of these concepts. Then test your formula on at least two pairs of values for a and b.

Solution for #11

Sample student response: If you substitute LCM(a, b) for P and GCF(a, b) for N in $P = a \div N \cdot b$, you get:

$$LCM(a,b) = a \div GCF(a,b) \cdot b$$

Sample tests:

GCF(15, 10) = 5	LCM(15, 10) = 30	$30 = 15 \div 5 \cdot 10$
GCF(9, 10) = 1	LCM(9, 10) = 90	$90 = 9 \div 1 \cdot 10$

Teacher's Note for #11. If students are having trouble understanding the question, refer them to Problems #3 and #7.

STAGE 3

Problem #12

12. Analyze a Paper Pool table with dimensions of 3.5 units and $4\frac{2}{3}$ units. Show how to find the path length (P), hits (H), and gap lengths (G). Sketch the path.

Questions and Conversations for #12

»   *What happens if you try to draw the path?* It's hard to draw the diagonal motion correctly because the grid doesn't guide you as well. Could it help to let each square on the paper represent something other than 1 unit?

»   *Are there grids with whole number dimensions in the same category as the one you are investigating?* Yes, there are. How can you calculate their dimensions?

Solution for #12

$P = 14$ diagonal units; $H = 7$; $G = 2\frac{1}{3}$ units. Sample strategy to find P:

»   Multiply a and b by 6 to make both of them whole numbers, 21 and 28.

»   Divide by 7 to get the dimensions of the smallest whole number grid, 4 by 3.

»   Find the length of the 4 by 3 path: $4 \cdot 3 = 12$

»   Find the length of the 28 by 21 path: $12 \cdot 7 = 84$

»   Find the length of the path on the original grid: $84 \div 6 = 14$

Sample strategies to find H and G:

»   The number of hits is the same as for the 4 by 3 grid. Add a and b: $4 + 3 = 7$.

»   To find the gap length, divide the perimeter of the original grid by the number of gaps: $16\frac{1}{3} \div 7 = 2\frac{1}{3}$.

Problem #13

13. Design a Paper Pool table whose path has gap lengths of 3.4 units and a path length of 25.5 units. What are the dimensions? How many hits will there be? Sketch the path. Explain your thinking process.

Questions and Conversations for #13

» *How can the connections between path length, gap length, greatest common factor, and least common multiple help you?* You might use these connections to help you "think backward."

Solution for #13

There are two solutions: $a = 1.7$, $b = 25.5$, $H = 16$ or $a = 5.1$, $b = 8.5$, $H = 8$.
The GCF is based on half the gap length, 1.7.
The LCM is based on the path length, 25.5.
Multiply both values by 10 to get whole numbers, 17 and 255.
Look for numbers that have a GCF of 17 and an LCM of 255.

$$a = 255 \text{ and } b = 17 \text{ or } a = 85 \text{ and } b = 51$$

Divide these numbers by 10 to find the dimensions of the tables we are searching for:

$$a = 25.5 \text{ and } b = 1.7 \text{ or } a = 8.5 \text{ and } b = 5.1$$

Divide the numbers by 17 (the GCF) to find the smallest whole-number grids in each category:

$$15 \text{ by } 1 \text{ or } 5 \text{ by } 3$$

This gives $15 + 1 = 16$ or $5 + 3 = 8$ hits.

Notice that in these pictures, the side of each square on the grid has a length of 1.7 units.

WRAP UP

Share Strategies

Students are likely to generate a greater variety of methods for solving Paper Pool problems than we were able to show in the "Solutions" section. Ask them to share and compare their strategies and observations.

Teacher's Note for Problem #13. If students share the strategies they developed for finding pairs of numbers that have a given GCF and LCM, you might mention Exploration 9: The GCF-LCM Connection, where they can investigate these ideas in more detail.

Summarize

Answer any remaining questions that students have. Summarize key points:

» Notice how organizing the data brought out many hidden patterns.
» Talk about why the path length is the least common multiple of a and b, and why the greatest common factor of a and b is a ratio of path lengths (and is also half the gap length).
» Discuss ratio patterns for pool tables belonging to the same category.

Further Exploration

Have students think of new questions to ask or ways to extend this exploration. Here are some possibilities:

» Can you use a and b to predict which pocket a ball will land in or what types of symmetry a path will have?
» At how many points will a path intersect itself?

Possible answers: $\dfrac{(a-1)\cdot(b-1)}{2}$ or $\dfrac{P-H+1}{2}$ for the smallest path in a category.

» For what pairs of values of H (hits) and P (path length) is there a path? What kinds of pairs have many paths?
» Is it possible to choose dimensions, a and b, so that the ball will travel forever (never hit a corner)?
» What happens if you use angles other than 45°?
» Does it make sense to consider a three-dimensional version of the Paper Pool problem?

Exploration 9

The GCF-LCM Connection

INTRODUCTION

Materials

- » Calculator (suggested)
- » Completed "Building Blocks Grid" (see http://www.routledge.com/Assets/ ClientPages/AdvancedCCMath.aspx)

Prior Knowledge

- » Exploration 1: Building Blocks
- » Exploration 5: How Many Factors?
- » Know the meaning of *greatest common factor* and *least common multiple*.

Learning Goals

- » Discover procedures for using prime factorizations to find factors and multiples.
- » Recognize prime numbers and calculate prime factorizations more fluently.
- » Develop and apply strategies for finding GCFs and LCMs using prime factorizations.
- » Discover and prove a connection between GCFs and LCMs.
- » Develop algebraic equations from computational strategies.
- » Communicate complex mathematical ideas clearly.
- » Persist in solving challenging problems.

Launching the Exploration

Motivation and purpose. You are about to use prime factorizations to look at greatest common factors and least common multiples in a whole new way. You will gain a deeper understanding of GCFs, LCMs, and exponents that will set the stage for future success in algebra.

Understanding the problem. Remind students that every natural number is a factor and a multiple of itself.

Review the connection between block diagrams and exponents: When prime factorizations are written in exponential form, the exponent on a base counts the number of blocks for that base. For example, the equation $1960 = 2^3 \cdot 5^1 \cdot 7^2$ shows that the block diagram of 1960 has three 2-blocks, one 5-block, and two 7-blocks.

 DOI: 10.4324/9781003232742-11

If a prime number does not appear in a prime factorization, you may include it by giving it an exponent of 0. The value of the base to the 0 power is equal to 1.

Look through the entire activity with students to help them see the big picture. In Stage 1, students use prime factorizations to develop new methods for calculating greatest common factors and least common multiples. In Stage 2, they extend these ideas to explore connections between GCFs and LCMs. Finally, in Stage 3, they investigate a new representation for prime factorizations and use it to solve problems involving GCFs and LCMs.

STUDENT HANDOUT

Stage 1

1. Write the prime factorizations of 30 and 12 as block diagrams. Describe a strategy to use the blocks to find the greatest common factor of 30 and 12. Explain your thinking.

2. Apply your method to find GCF(270, 110) and GCF(132, 455).

3. List five multiples of 45 and show each prime factorization using a block diagram. How can you use the blocks to recognize each number as a multiple of 45? Explain.

4. Describe a strategy to use the blocks to find the least common multiple of 270 and 110. Explain your thinking. (You may use your block diagrams from Problem #2.)

5. Apply your strategy to find LCM(132, 455) and LCM(155, 2015). What interesting things do you observe? Why do these things happen?

6. Write the prime factorizations of 30 and 12 in exponential form. Create and describe processes for using this representation to find GCF(30, 12) and LCM(30, 12).

7. With words or a formula, explain how to use a, b, and $GCF(a, b)$ to calculate $LCM(a, b)$. Use block diagrams to explain why the process works.

8. Test your process or formula from Problem #7 on at least two pairs of numbers.

9. How can you use block diagrams to find the GCF of three or more numbers? Explain using at least one example.

10. How can you use block diagrams to find the LCM of three or more numbers? Describe the process using at least one example. Does your formula still apply? Explain.

11. What is this Venn diagram about? How does it work? How can you use it to do calculations with the numbers 120 and 700?

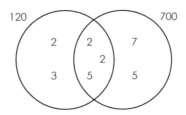

12. Use the Venn diagram to find at least two more pairs of numbers that have the same greatest common factor and least common multiple as 120 and 700. Explain your thinking.

13. Show how you can use a Venn diagram to find *all* pairs of numbers a, b with $GCF(a, b) = 7$ and $LCM(a, b) = 84$. What happens when you apply your method to $GCF(a, b) = 14$ and $LCM(a, b) = 21$? Why?

14. Draw a three-set Venn diagram showing the numbers 630, 280, and 546. Explain how to use it to find $GCF(630, 280, 546)$ and $LCM(630, 280, 546)$.

15. Use your Venn diagrams to help you create a formula for $LCM(a, b, c)$ using a, b, c, and greatest common factors as variables. Explain your thinking process and tell why your formula works. Then test it on the numbers from Problem #14.

TEACHER'S GUIDE

STAGE 1

Problem #1

1. Write the prime factorizations of 30 and 12 as block diagrams. Describe a strategy to use the blocks to find the greatest common factor of 30 and 12. Explain your thinking.

Questions and Conversations for #1

This section contains ideas for conversations, mainly in the form of questions that students may ask or that you may pose to them. Be sure to allow students to do most of the thinking and talking!

> » *How can you recognize factors of block diagrams?* A factor of a block diagram is contained within the block diagram.

Teacher's Note. In this exploration, when I say "a factor of a block diagram," it really means "a factor of the number represented by a block diagram." My phrasing isn't quite as precise, but it makes some sentences much easier to read!

> » *Is the greatest common factor of a number the same as its greatest common prime factor?* Not generally. For example, the greatest common factor of 18 and 12 is 6, but their greatest common *prime* factor is 3!

Solution for #1

The block diagrams for 30 and 12 look like this, with the common prime factors shaded.

$$\boxed{2}\boxed{3}\boxed{5} \qquad \boxed{2}\boxed{2}\boxed{3}$$

GCF(30, 12) = 6 because the common prime factors are 2 and 3, and this represents the number 6.

Problem #2

2. Apply your method to find GCF(270, 110) and GCF(132, 455).

Questions and Conversations for #2

> » *Does every pair of natural numbers have a GCF?* Yes. All natural numbers share a certain common factor. What is it?

Solution for #2

GCF(270, 110) = 10 because the common prime factors, 2 and 5, represent the number 10.

| 2 | 3 | 3 | 3 | 5 | | 2 | 5 | 11 |

GCF(132, 455) = 1 because there are no blocks (prime factors) in common. (The Building Blocks grid shows that "no blocks" represents the number 1.)

| 2 | 2 | 3 | 11 | | 5 | 7 | 13 |

Problem #3

3. List five multiples of 45 and show each prime factorization using a block diagram. How can you use the blocks to recognize each number as a multiple of 45? Explain.

Questions and Conversations for #3

» *What do all of the block diagrams for the multiples of 45 have in common?* They have a certain collection of blocks in common. What is it?

Solution for #3

We show the first five multiples. Students may have chosen others.

45	90	135	180	225
3 | 3 | 5	2 | 3 | 3 | 5	3 | 3 | 3 | 5	2 | 2 | 3 | 3 | 5	3 | 3 | 5 | 5

The pictures are shaded to show that the block diagram of 45 is contained inside the block diagrams of all of its multiples (including itself).

Problem #4

4. Describe a strategy to use the blocks to find the least common multiple of 270 and 110. Explain your thinking. (You may use your block diagrams from Problem #2.)

Questions and Conversations for #4

» *Is it possible to start with one block diagram and change it into the LCM?* Yes, this is one strategy. Think of the blocks that are "missing" from the diagram that you start with.

» *Is it possible find the LCM by focusing on one prime factor at a time?* Yes. This is a second strategy. How can you decide how many of each factor you need?

» *What happens if you just join the block diagrams of the two original numbers?* The answer is a multiple of both numbers but it is not necessarily the smallest possible common multiple.

»   *Did you notice any "extra" blocks in the process of joining the diagrams?* You might see blocks that are not needed when you join diagrams. What can you do with them? Do they remind you of something you've seen before?

Teacher's Note. Encourage students to check potential LCMs by asking:
- Is the block diagram a multiple of both numbers? How can you tell?
- Is the block diagram the smallest multiple of both numbers? How can you tell?

Solution for #4

The answer is 2970:

| 2 | 3 | 3 | 3 | 5 | 11 |

There are three common strategies.

Strategy 1: Begin with one block diagram and join it to every prime factor that is missing from the other one. Example: Begin with 270 and attach an 11-block because it is the only block 110 has that 270 is missing.

| 2 | 3 | 3 | 3 | 5 | ⟶   ⟵ | 11 |

Strategy 2: For each prime factor, choose the larger number of blocks possessed by either number and then attach all of these. Example: The 2-, 5-, and 11-blocks appear at most once in either number separately. The 3-block appears at most three times. Join these to form the LCM.

| 2 | ⟶   ⟵ | 3 | 3 | 3 | ⟶   ⟵ | 5 | ⟶   ⟵ | 11 |

Strategy 3: Eliminate the common blocks (the GCF) from either diagram and then join the remaining blocks to the other diagram. Example: Begin with 270 and get rid of the 2- and 5-blocks that the numbers have in common. Then attach the remaining three 3-blocks to 110's diagram.

| ⊠ | 3 | 3 | 3 | ⊠ | ⟶   ⟵ | 2 | 5 | 11 |

Teacher's Note. Students can check the final result by making sure it contains both original block diagrams and that all of the blocks are needed. If we eliminate any of them, it will fail to contain one or both of the original numbers.

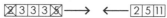

| 2 | 3 | 3 | 3 | 5 | 11 | | 2 | 3 | 3 | 3 | 5 | 11 |

Problem #5

5. Apply your strategy to find LCM(132, 455) and LCM(155, 2015). What interesting things do you observe? Why do these things happen?

Solution for #5

Sample student solutions:
LCM(132, 455) = 60,060
The two prime factorizations are:

| 2 | 2 | 3 | 11 | | 5 | 7 | 13 |

Because they have no common prime factors, there are no blocks to eliminate. We join the two diagrams to get $2 \cdot 2 \cdot 3 \cdot 11 \cdot 5 \cdot 7 \cdot 13 = 60,060$.

This shows that if the GCF of two numbers is 1, you can find the LCM just by multiplying the numbers.

LCM(155, 2015) = 2015
The two prime factorizations are:

| 5 | 31 | | 5 | 13 | 31 |

If you eliminate the common blocks (5 and 31) from either one and join it to the other, you will get a LCM of $5 \cdot 13 \cdot 31 = 2015$.

This suggests that if one number is a factor (or multiple) of the other, the smaller number is the GCF and the larger number is the LCM. Encourage students to explain why this happens!

Problem #6

6. Write the prime factorizations of 30 and 12 in exponential form. Create and describe processes for using this representation to find GCF(30, 12) and LCM(30, 12).

Questions and Conversations for #6

» *How is this different than Problems #1 and #4?* You are being asked to use a different representation of the prime factorization—the exponential form instead of the block diagram.

» *Can you use your strategies from Problems #1 and #4 as guides to creating processes for the exponential form?* Possibly—depending on the strategies you used. If not, consider looking at one prime factor at a time.

Solution for #6

GCF(30, 12)
To find all blocks that two diagrams have in common, look at one prime factor at a time and choose the expression with the smaller of the two exponents. Example:

$$30 = 2^1 \cdot 3^1 \cdot 5^1 \qquad\qquad 12 = 2^2 \cdot 3^1$$

For 2, we choose 2^1. For 3, both exponents are equal, so we have 3^1. And for 5, we choose 5^0 (because we can write 12 as $2^2 \cdot 3^1 \cdot 5^0$).
Putting all of this together, we get $2^1 \cdot 3^1 \cdot 5^0 = 2 \cdot 3 \cdot 1 = 6$.

LCM(30, 12)
To find the smallest block diagram that contains both numbers, we choose the expression with the greater (or equal) exponent for each factor: 2^2, 3^1 and 5^1.
Multiplying these gives us $2^2 \cdot 3^1 \cdot 5^1 = 4 \cdot 3 \cdot 5 = 60$.

STAGE 2

Problem #7

7. With words or a formula, explain how to use a, b, and GCF(a, b) to calculate LCM(a, b). Use block diagrams to explain why the process works.

Questions and Conversations for #7

» *Joining and taking away blocks represent which operations?* Joining blocks represents multiplication. Taking away blocks means division. (It is important to remind ourselves of this, because we are accustomed to think of joining as standing for addition and taking away as standing for subtraction.)

Teacher's Note. As you read this solution, remember that joining blocks represents multiplication and taking them away represents division.

Solution for #7

Suppose you begin with a. Taking away the "common blocks" means dividing by the greatest common factor of a and b.

$$a \div GCF(a,b)$$

Joining this to the other block diagram represents multiplying by b.

$$LCM(a,b) = a \div GCF(a,b) \cdot b$$

If students manipulate the blocks in a different order, they may get

$$LCM(a,b) = b \div GCF(a,b) \cdot a \ \text{ or } \ LCM(a,b) = a \cdot b \div GCF(a,b)$$

Problem #8

8. Test your process or formula from Problem #7 on at least two pairs of numbers.

Teacher's Note for #8. Encourage students to choose numbers for which they can find the GCF and LCM fairly easily without their formula or process so that they can use their process to check that it gives the same results.

Solution for #8

Sample solutions:

$a = 6; b = 8$ $\qquad$ GCF$(6, 8) = 2$ $\qquad$ LCM$(6, 8) = 24$

$$a \div GCF(a, b) \cdot b = 6 \div 2 \cdot 8 = 3 \cdot 8 = 24$$

$a = 12; b = 9$ $\qquad$ GCF$(12, 9) = 3$ $\qquad$ LCM$(12, 9) = 36$

$$a \div GCF(a, b) \cdot b = 12 \div 3 \cdot 9 = 4 \cdot 9 = 36$$

Problem #9

9. How can you use block diagrams to find the GCF of three or more numbers? Explain using at least one example.

Questions and Conversations for #9

» *Does it make sense to try to think of all three numbers at once, rather than two at a time?* Yes, it can. In fact, this might be easier.

Solution for #9

Find the blocks that are contained in all three numbers' block diagrams. For example: GCF(189, 6, 441) = 3 because the following diagrams have a common factor of only 3 (even though some pairs have more in common).

| 3 | 3 | 3 | 7 | $\quad$ | 2 | 3 | $\quad$ | 3 | 3 | 7 | 7 |

Problem #10

10. How can you use block diagrams to find the LCM of three or more numbers? Describe the process using at least one example. Does your formula still apply? Explain.

Questions and Conversations for #10

» *Do the strategies for finding the LCM of two numbers still work for three numbers?* Some do. Others may not, or may need to be adjusted. To stay on the right track, remain focused on the meaning of *least common multiple*—the smallest number that is a multiple of all three numbers.

Solution for #10

Sample solution: For each prime factor, find the block diagram in which it appears the most often, and select the blocks for that factor. Join these to produce the LCM. Example: LCM(189, 6, 441) = 2646 because you can choose the 2 from the second diagram, three 3s from the left one, and two 7s from the one on the right. Then put these together to get $2 \cdot 3 \cdot 3 \cdot 3 \cdot 7 \cdot 7 = 2646$.

| 2 | 3 | 3 | 3 | 7 | 7 |

This makes sense because it is the smallest collection of blocks (prime factors) that contains all three block diagrams above. The most natural way to adjust the formula for three numbers does not work:

$$a \cdot b \cdot c \div GCF(a,b,c) = 189 \cdot 6 \cdot 441 \div 3 = 166,698$$

Teacher's Note. See Problem #15 to search for a formula that does work!

STAGE 3

Problem #11

11. What is this Venn diagram about? How does it work? How can you use it to do calculations with the numbers 120 and 700?

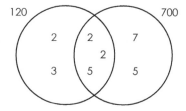

Questions and Conversations for #11

» *How can you connect the Venn diagram to the block diagrams?* Because the numbers inside the Venn diagrams are prime numbers, you might imagine them as blocks. Blocks that belong to the same set in the Venn diagram are multiplied.

» *What quantity do the common members of the two sets represent? (This is called the intersection of the sets.)* The intersection of the sets represents a greatest common factor. Why?

» *What quantity do the combined members of the two sets represent? (This is called the union of the sets.)* The union of the sets represents a least common multiple. Why?

Solution for #11

Each circle contains the prime factorization of its associated number.

GCF(120, 700) = $2 \cdot 2 \cdot 5 = 20$ shows up in the *intersection* (overlap) of the circles.

LCM(120, 700) = $2 \cdot 2 \cdot 2 \cdot 3 \cdot 5 \cdot 5 \cdot 7 = 4200$ appears as the *union* (combination) of the two circles. Notice how the factors in the GCF are counted just once.

Problem #12

12. Use the Venn diagram to find at least two more pairs of numbers that have the same greatest common factor and least common multiple as 120 and 700. Explain your thinking.

Questions and Conversations for #12

See Questions and Conversations for #11.

Solution for #12

The numbers 2, 2, and 5 must remain in the middle to preserve the GCF, while the seven numbers inside the circles must not change so that the LCM stays the same.

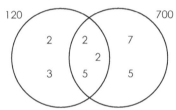

This still allows you to move around the 2, 3, 5, and 7 on the sides! For example, if we move the 3 to the right set, the two numbers will be 40 and 2100.

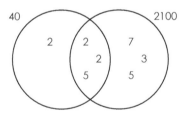

Now, if you then move the 7 to the left set, the two numbers will be 280 and 300.

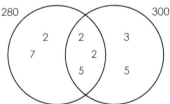

Other pairs that you can find this way are 20 and 4200, 140 and 600, 200 and 420, 60 and 1400, and 100 and 840.

Problem #13

13. Show how you can use a Venn diagram to find *all* pairs of numbers *a, b* with GCF(*a, b*) = 7 and LCM(*a, b*) = 84. What happens when you apply your method to GCF(*a, b*) = 14 and LCM(*a, b*) = 21? Why?

Questions and Conversations for #13

» *What has to stay the same in the Venn diagram so that the GCF does not change?* The numbers in the intersection of the two sets cannot change. Why not?
» *What has to stay the same in the Venn diagram so that the LCM is not affected?* The union of the two sets cannot change. Why not?
» *What can change?* Sometimes you can move prime factors from one set to another. When is it okay to do this? When is it not okay?

Teacher's Note. The words *union* and *intersection* are included here to make the sentences simpler. Using these words with your students is optional.

Solution for #13

The pair 21, 28 is one possibility:

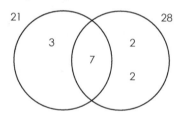

You can put also put the 2, 2, and 3 together on either side to get the pair 7, 84. This is the only other possibility. (Why can't you move a 2 into the left set?)

You can't create a Venn diagram for GCF(*a, b*) = 14 and LCM(*a, b*) = 21. The intersection contains 2 and 7 due to the GCF of 14. But the LCM is 21, and 2 is not a factor of 21, so the Venn diagram cannot contain the number 2! The real problem is that the LCM must be a multiple of the GCF.

Problem #14

14. Draw a three-set Venn diagram showing the numbers 630, 280, and 546. Explain how to use it to find GCF(630, 280, 546) and LCM(630, 280, 546).

Questions and Conversations for #14

» How should you draw your Venn diagram so that it contains regions that represent all possible combinations of members of the three sets?

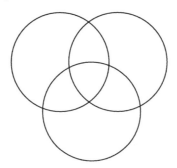

Solution for #14

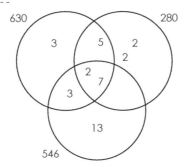

GCF(630, 280, 546) = 14 because only 2 and 7 are common to all three numbers.

LCM(630, 280, 546) = 32,760, the product of each number in the diagram:

$$2 \cdot 2 \cdot 2 \cdot 3 \cdot 3 \cdot 5 \cdot 7 \cdot 13 = 32,760$$

Problem #15

15. Use your Venn diagrams to help you create a formula for LCM(*a*, *b*, *c*) using *a*, *b*, *c*, and greatest common factors as variables. Explain your thinking process and tell why your formula works. Then test it on the numbers from Problem #14.

Questions and Conversations for #15

» *Based on your Venn diagram, what different greatest common factors might be a part of your formula?* The formula might involve GCF(*a*, *b*), GCF(*a*, *c*), GCF(*b*, *c*), or GCF(*a*, *b*, *c*) because each of these has a region representing it in the Venn diagram.

Solution for #15

$$LCM(a,b,c) = a \cdot b \cdot c \div GCF(a,b) \div GCF(a,c) \div GCF(b,c) \cdot GCF(a,b,c)$$

The first three GCFs correct the double-count of factors in the two-set over-laps, and the final GCF ensures that the 2 and 7 in the very middle are included exactly once. (This may take quite a bit of thought, so don't expect your students—or yourself—to understand it right away!) If you test this on the numbers from Problem #14, you get:

$$630 \cdot 280 \cdot 546 \div 70 \div 42 \div 14 \cdot 14 = 32,760$$

WRAP UP

Share Strategies

Ask students to share their strategies for using block diagrams and exponents to find GCFs and LCMs of numbers. Have them explain why the strategies make sense.

Summarize

Answer any remaining questions students have. Summarize these key points:

» Prime factorizations are helpful tools for finding greatest common factors and least common multiples. There are many strategies for using them to find LCMs.

» There is a connection between greatest common factors and least common multiples. The connection can be described in words or formulas. One way to find the LCM of a pair of numbers is to multiply them and divide by their GCF. If you change the order of the process and divide one of the numbers by the GCF before multiplying by the other number, the calculations are often simpler.

» If two numbers have a GCF of 1, their LCM is equal to their product.

» You can use Venn diagrams to represent relationships between prime factorizations.

Further Exploration

Have students think of new questions to ask or ways to extend this exploration. Here are some possibilities.

» Use Venn diagrams to explore GCFs and LCMs of more than three numbers. Keep in mind that it can be quite challenging to make Venn diagrams of four or more sets. (The difficulty lies in representing all possible ways that the sets can intersect.)

» Extend Problem #15 by looking for formulas involving greatest common factors and least common multiples of four and more numbers. Find patterns between the formulas.

» Use ideas from this activity to explore GCFs and LCMs of algebraic expressions.

Exploration 10

Mathematical Mystery Code

INTRODUCTION

Prior Knowledge

» Know the meaning of *prime factorizations* and have strategies for calculating them.

» Understand whole number exponents.
  Note. Stage 3 may work best if students do not yet know about negative exponents.

Learning Goals

» Generalize understanding of the place value concept.
» Develop strategies for solving a problem based on prime factorizations.
» Analyze and extend prime factorization patterns.
» Use prime factorizations to reason about properties of exponents.
» Describe numeric patterns with words and equations.
» Communicate complex mathematical ideas clearly.
» Persist in solving challenging problems.

Launching the Exploration

Motivation and purpose. To students: There's a reason this is called the Mathematical Mystery Code—you won't even know what it's really about until you've worked on it for a while! Be patient—it could take at least a few days to a week to "crack the code." You are receiving only the handout for Stage 1 right now, because the remaining pages give some things away. Happy decoding!

Understanding the problem. Tell students that it is possible to solve parts of the code correctly without figuring out the whole thing. They are many "small" patterns inside a larger general pattern that ties all of them together.

After students have completed Stage 1, look through the rest of the exploration with them to see the big picture. In Stage 2, they focus on understanding why the code works, which leads to an exploration of properties of exponents. The most important (and possibly the most challenging) problem is #7. In Stage 3, the emphasis shifts to applying their new knowledge to solve problems about a new kind of "magic square." This offers students a chance to begin thinking about the meaning of negative exponents.

> **Teacher's Note.** Hand out only Stage 1 of the activity at first. Once students complete it, you can give them the remaining pages.

DOI: 10.4324/9781003232742-12

STUDENT HANDOUT

Stage 1

1. Figure out as much as you can about this code. Then find the codes for the numbers 16–30. Describe your discoveries and tell how you made them.

Original Number	Code Number	Original Number	Code Number
0	none	16	
1	0	17	
2	1	18	
3	10	19	
4	2	20	
5	100	21	
6	11	22	
7	1,000	23	
8	3	24	
9	20	25	
10	101	26	
11	10,000	27	
12	12	28	
13	100,000	29	
14	1,001	30	
15	110		

2. Create and complete this table for the natural numbers 2–20. Use the example for the number 12 as a guide. Begin by writing the prime factorization of each number in the second column. Then use it to find a way to complete the columns on the right.

Natural Number	Prime Factorization	19	17	13	11	7	5	3	2
...									
12	$2 \cdot 2 \cdot 3$	0	0	0	0	0	0	1	2
...									

3. Compare the table with the code numerals. What do you notice? Why do the codes of prime numbers look like powers of 10? Use your table from Problem #2 to explain.

4. Copy and complete this table. The bases of your exponential expressions must be prime numbers.

Number Sentence	Prime Factorizations	Code
$25 \cdot 5 = 125$	$5^2 \cdot 5^1 = 5^3$	$200 + 100 = 300$
$8 \cdot 8 = 64$		
$243 \cdot 9 =$ ____		
	$7^1 \cdot 7^3 =$ ___	
	$3^3 \cdot$ ___ $= 3^3$	
		$20 + 20 = 40$
		$5 + 0 = 5$

5. Use words or an equation to describe a pattern in the middle column. Be as complete and clear as you can. If you use words, include the vocabulary *base*, *exponent*, *factor*, and *product*.

6. What are the values of 2^0 and 3^0? What about other expressions with exponents of 0? How do you know? Use the Mathematical Mystery Code and the table in Problem #4 to help you explain.

7. Explain why the code for every counting number must be the sum of the codes of its factors.

This is a *magic square*. Each row, column, and main diagonal has the same sum (15).

2	7	6
9	5	1
4	3	8

8. Use this square to create a magic square based on multiplication instead of addition. Each row, column, and main diagonal will have the same product instead of the same sum. Explain your thinking and tell why your method works. *Note.* The numbers in your "multiplication" square will still be natural numbers, but they will not be consecutive.

9. Subtract 1 from each number in the original magic square to make a new one. How much did its sum change? Why? Use this new "addition" magic square to create another one based on multiplication. How does the new product compare to the original one? Why?

10. Continue this process of subtracting one and creating new "multiplication" squares. Again, describe how the sums and products of the squares change.

11. What does your experience with "multiplication" magic squares teach you about negative exponents? Describe what you have learned about them—what they mean and why.

TEACHER'S GUIDE

STAGE 1

Problem #1

1. Figure out as much as you can about this code. Then find the codes for the numbers 16–30. Describe your discoveries and tell how you made them.

Original Number	Code Number
0	none
1	0
2	1
3	10
4	2
5	100
6	11
7	1,000
8	3
9	20
10	101
11	10,000
12	12
13	100,000
14	1,001
15	110

Questions and Conversations for #1

This section contains ideas for conversations, mainly in the form of questions that students may ask or that you may pose to them. Be sure to allow students to do most of the thinking and talking!

» *Do all code numbers contain only the digits 0, 1, 2, and 3?* No. They can contain any digit.

> **Teacher's Note.** Be sure to allow students plenty of time to think on their own (or with partners) before you begin to discuss these questions. Students often start by looking only at patterns of change in the code numbers themselves, ignoring original numbers. These patterns tend to fall apart eventually. After they've explored this for a while, suggest that they begin to focus on how the original numbers relate to the code numbers.

» *What (original) numbers have code numbers of 0, 1, 2, and 3?* 1, 2, 4, and 8. Is there a pattern? Can you make any predictions?

» *What numbers have code numbers of 10 and 20?* 3 and 9. If you compare this to the previous question, do you see another pattern? Can you make any predictions?

» *What numbers have codes that look like powers of 10?* 2, 3, 5, 7, 11, etc. What can you say about these numbers?

» *What do you notice about the code numbers for 2, 3, and 6?* Look at how the numbers 2, 3, and 6 are related. Then look at how their code numbers are related.

» *What happens to the code number when you double the original number? When you triple it?* When you double a number, its code number increases by 1. When you triple it, its code number increases by 10.

Solution for #1

Original Number	Code Number
16	4
17	1,000,000
18	21
19	10,000,000
20	102
21	1010
22	10,001
23	100,000,000
24	13
25	200
26	100,001
27	30
28	1002
29	1,000,000,000
30	111

The prime numbers look like successive powers of 10 when you put them into code, beginning with the number 2, which has a code number of 1. To find the code of a composite number, factor it in any way you like, then add the code numbers of each factor. For example, to find the code number of 24, you could factor it as $6 \cdot 4$. Then find the code numbers of 6 and 4 and add them: $11 + 2 = 13$.

It doesn't matter what factors you choose. You can even use three or more factors. For instance, $2 \cdot 2 \cdot 2 \cdot 3$ results in $1 + 1 + 1 + 10 = 13$.

STAGE 2

Problem #2

2. Create and complete this table for the natural numbers 2–20. Use the example for the number 12 as a guide. Begin by writing the prime factorization of each number in the second column. Then use it to find a way to complete the columns on the right.

Natural Number	Prime Factorization	19	17	13	11	7	5	3	2
...									
12	$2 \cdot 2 \cdot 3$	0	0	0	0	0	0	1	2
...									

Questions and Conversations for #2

» *What do the numbers in the columns at the right mean?* In the example for the number 12, the 1 in the column labeled 3 and the 2 in the column labeled 2 mean that the prime factorization of 12 contains one factor of 3 and two factors of 2. The remaining columns show 0 because these numbers are not in the prime factorization of 12.

» *Some people might describe this as a place value code. Why?* Think of each word—place and value—individually. The value of each digit depends upon its place in the numeral. For example, in the code number 22, the left digit has a value of 9 (two factors of 3, or 3^2), while the right digit has a value of 4 (two factors of 2, or 2^2).

» *Why is "none" written as the code number for 0?* The number 0 cannot be put into code, because it does not have a prime factorization. (Also, 0 cannot be written in exponential form.)

Solution for #2

Natural Number	Prime Factorization	19	17	13	11	7	5	3	2
2	2	0	0	0	0	0	0	0	1
3	3	0	0	0	0	0	0	1	0
4	$2 \cdot 2$	0	0	0	0	0	0	0	2
5	5	0	0	0	0	0	1	0	0
6	$2 \cdot 3$	0	0	0	0	0	0	1	1
7	7	0	0	0	0	1	0	0	0
8	$2 \cdot 2 \cdot 2$	0	0	0	0	0	0	0	3

Natural Number	Prime Factorization	19	17	13	11	7	5	3	2
9	3·3	0	0	0	0	0	0	2	0
10	2·5	0	0	0	0	0	1	0	1
11	11	0	0	0	1	0	0	0	0
12	2·2·3	0	0	0	0	0	0	1	2
13	13	0	0	1	0	0	0	0	0
14	2·7	0	0	0	0	1	0	0	1
15	3·5	0	0	0	0	0	1	1	0
16	2·2·2·2	0	0	0	0	0	0	0	4
17	17	0	1	0	0	0	0	0	0
18	2·3·3	0	0	0	0	0	0	2	1
19	19	1	0	0	0	0	0	0	0
20	2·2·5	0	0	0	0	0	1	0	2

Problem #3

3. Compare the table with the code numerals. What do you notice? Why do the codes of prime numbers look like powers of 10? Use your table from Problem #2 to explain.

Questions and Conversations for #3

» *Why are the prime numbers listed in the reverse order at the top of the table?* This causes the numbers in the columns below to match the digits in the code numbers in the same order that they actually appear in the code.

Solution for #3

If you ignore the 0s on the left (shown in grey), then the numbers in the right part of the table exactly match the codes!

The codes of prime numbers look like powers of 10 because the prime factorization of a prime number is just the number itself. This means it appears 1 time in its own prime factorization and all prime numbers less than it show up 0 times. This creates a code numeral of 1 followed by a string of zeros.

Problem #4

4. Copy and complete this table. The bases of your exponential expressions must be prime numbers.

Number Sentence	Prime Factorizations	Code
$25 \cdot 5 = 125$	$5^2 \cdot 5^1 = 5^3$	$200 + 100 = 300$
$8 \cdot 8 = 64$		

Number Sentence	Prime Factorizations	Code
$243 \cdot 9 =$ ____		
	$7^1 \cdot 7^3 =$ ___	
	$3^3 \cdot$ ___ $= 3^3$	
		$20 + 20 = 40$
		$5 + 0 = 5$

Questions and Conversations for #4

>> *What is special about all of the numbers that appear in the left column of the table?* They are all powers of prime numbers.

>> *Within a row, how do the equations in the three columns relate?* They all represent the same number sentence. You can translate the equation in any column into the equation in any other column. Each row is focused on a single prime number.

Solution for #4

Number Sentence	Prime Factorizations	Code
$25 \cdot 5 = 125$	$5^2 \cdot 5^1 = 5^3$	$200 + 100 = 300$
$8 \cdot 8 = 64$	$2^3 \cdot 2^3 = 2^6$	$3 + 3 = 6$
$243 \cdot 9 = 2187$	$3^5 \cdot 3^2 = 3^7$	$50 + 20 = 70$
$7 \cdot 343 = 2401$	$7^1 \cdot 7^3 = 7^4$	$1000 + 3000 = 4000$
$27 \cdot 1 = 27$	$3^3 \cdot 3^0 = 3^3$	$30 + 0 = 30$
$9 \cdot 9 = 81$	$3^2 \cdot 3^2 = 3^4$	$20 + 20 = 40$
$32 \cdot 1 = 32$	$2^5 \cdot 2^0 = 2^5$	$5 + 0 = 5$

Problem #5

5. Use words or an equation to describe a pattern in the middle column. Be as complete and clear as you can. If you use words, include the vocabulary *base*, *exponent*, *factor*, and *product*.

Questions and Conversations for #5

>> *What features do the equations have in common?* The base of every exponential expression in an equation is the same. The exponent of the product can be predicted from the exponents of the factors.

Solution for #5

When the bases in the factors are the same, the base in the product is also the same. The exponent of the product is the sum of the exponents in the factors.

In algebraic form, you could write this as $a^m \cdot a^n = a^{m+n}$.

Problem #6

6. What are the values of 2^0 and 3^0? What about other expressions with exponents of 0? How do you know? Use the Mathematical Mystery Code and the table in Problem #4 to help you explain.

Questions and Conversations for #6

» *Can you find the expressions 2^0 and 3^0 in the table from Problem #4?* Yes. Look at the middle column of the fifth and seventh rows. The values of the expressions appear in the left column. What do you think about other bases to the power of 0?

Solution for #6

The last row of the table in Problem #4 shows that $2^0 = 1$. We can think of 2^0 in the equation $2^5 \cdot 2^0 = 2^5$ as "zero factors of 2." For the equation $32 \cdot \underline{} = 32$ to be true, the value of 2^0 must be 1. We can use the same argument with 3^0 in the fifth row of the table, so $3^0 = 1$ as well.

We could draw the same conclusion no matter what prime number we use as a base. In fact, if a is any natural number, $a^0 = 1$.

Problem #7

7. Explain why the code for every counting number must be the sum of the codes of its factors.

Questions and Conversations for #7

» *What does "the sum of the codes of its factors" mean?* You are essentially being asked to explain why you add the code numbers. Break the phrase down into its parts: choose a number, find a pair of its factors, find their code numbers, and add them. This diagram might help you to picture it.

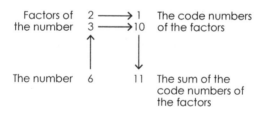

> » *What does it mean to explain why the code works this way?* Think of the code as a way to represent the original number's prime factorization. Use this idea to explain why the code numbers must add the way they do.
> » *Does the reason have anything to do with the pattern from Problem #5?* Yes, because Problem #5 shows the connection between multiplication and addition.

Solution for #7

When you add the code numbers, you are just counting the times each prime number shows up in the prime factorizations of the original numbers.

Let's use $50 \cdot 12 = 600$ as an example. The code numbers are 201, 12, and 213.

$50 = 5^2 \cdot 3^0 \cdot 2^1$	2 factors of 5	0 factors of 3	1 factor of 2
$12 = 5^0 \cdot 3^1 \cdot 2^2$	0 factors of 5	1 factor of 3	2 factors of 2
$600 = 5^2 \cdot 3^1 \cdot 2^3$	2 factors of 5	1 factor of 3	3 factors of 2

Adding digits in the code numbers is just like adding exponents in the prime factorizations!

STAGE 3

This is a *magic square*. Each row, column, and main diagonal has the same sum (15).

2	7	6
9	5	1
4	3	8

Problem #8

8. Use this square to create a magic square based on multiplication instead of addition. Each row, column, and main diagonal will have the same product instead of the same sum. Explain your thinking and tell why your method works. *Note.* The numbers in your "multiplication" square will still be natural numbers, but they will not be consecutive.

> **Teacher's Note for #8.** Give students plenty of time to develop their own strategies for finding a "multiplication" magic square. If they continue to struggle, point out that they are looking for connections between multiplication and addition. Ask them where else in the exploration they have seen this.

Solution for #8

Think of each number in the magic square as a code number. Replace it with its "original number." This is the same as taking 2 to the power of each number in the square.

2^2	2^7	2^6
2^9	2^5	2^1
2^4	2^3	2^8

4	128	64
512	32	2
16	8	256

This works because the original numbers now just count the factors of 2. Every row, column, and main diagonal has 15 factors of 2. Because the original sum was 15, the product is $2^{15} = 32,768$.

Of course, you could also use a number other than 2 as a base!

Problem #9

9. Subtract 1 from each number in the original magic square to make a new one. How much did its sum change? Why? Use this new "addition" magic square to create another one based on multiplication. How does the new product compare to the original one? Why?

Questions and Conversations for #9

» *Is it okay to have negative numbers and fractions in the magic squares?* Yes! This will eventually happen as you continue subtracting one from the numbers and changing the addition squares into multiplication squares.

» *How many times should you subtract one and then make new multiplication squares?* It's up to you. You should probably continue at least until you feel that you are able to answer the final question about negative exponents.

Solution for #9

The new sum is 12. This decreased by 3 because each entry in a row, column, or diagonal decreased by 1.

The new product is $2^{12} = 4096$. It became $\dfrac{1}{8}$ as large because lowering each exponent by 1 made each number half as large. Doing this three times divided the product by $2^3 = 8$.

1	6	5
8	4	0
3	2	7

2	64	32
256	16	1
8	4	128

Problem #10

10. Continue this process of subtracting one and creating new "multiplication" squares. Again, describe how the sums and products of the squares change.

> **Teacher's Note.** We can see again why $2^0 = 1$. The number in the original multiplication square was 2. When we decreased the exponent by 1, we made the number half as large, and it became 1.

Questions and Conversations for #10

See Questions and Conversations for #9.

Solution for #10

The sums of the next two "addition" squares are 9 and 6. They continue to decrease by 3 each time.

The products of the next two "multiplication" squares are 512 and 64. They continue to get $\frac{1}{8}$ as large each time.

0	5	4
7	3	-1
2	1	6

1	32	16
128	8	$\frac{1}{2}$
4	2	64

-1	4	3
6	2	-2
1	0	5

$\frac{1}{2}$	16	8
64	4	$\frac{1}{4}$
2	1	32

Continuing to subtract and divide has created some negative numbers and fractions in the squares.

Problem #11

11. What does your experience with "multiplication" magic squares teach you about negative exponents? Describe what you have learned about them—what they mean and why.

Solution for #11

Because the numbers in the "addition" magic squares represent exponents in the "multiplication" squares, the results suggest that $2^{-1} = \frac{1}{2}$ and $2^{-2} = \frac{1}{4}$. The negative exponents create fractional values.

Some students may observe more:

» When you write 2^{-2} as a fraction, its denominator is equal to 2^2. In general, you can write a^{-n} in fraction form by making the numerator equal to 1 and the denominator to a^n. (We are thinking of n as a natural number.)

» If students are familiar with *reciprocals*, they might observe that making an exponent negative causes the value of the exponential expression to become the reciprocal of its original value.

151

WRAP UP

Share Strategies

Ask students to share their original strategies for solving the code.

Summarize

Answer any remaining questions that students have. Summarize these points:

» Place value refers to situations where the value of a digit depends on its place in the numeral. There can be different kinds of place value.

» The mystery code is a representation for prime factorizations.

» Exponential expressions satisfy properties that may be stated algebraically as

$$a^m \cdot a^n = a^{m+n}$$
$$a^0 = 1$$
$$a^{-n} = \frac{1}{a^n}$$

» The Mathematical Mystery Code helps you understand reasons for these properties.

Further Exploration

Ask students to think of ways to continue or extend this exploration. Here are some possibilities:

» What is the connection between the Mathematical Mystery Code and the colored block diagrams from Exploration 1?

» The Mystery Code eventually breaks down. What is the smallest natural number that can't be put into code? Why does this happen? How could you fix this problem? Remember that 0 is not a natural number. (The answer to the first question is 1024. The problem is that its prime factorization has 10 twos and 10 is a two-digit number.)

» What happens to the codes when you divide numbers? How could you use the code to represent fractions? How does all of this relate to properties of exponents?

» What happens to the code when you square a number? What happens when you take the square root? How does this relate to properties of exponents? How can you extend these ideas to cubes and cube roots?

» How can you create "multiplication" magic squares whose numbers are not just powers of a single base?

References

Burkhart, J. (2009). Building numbers from primes. *Mathematics Teaching in the Middle School, 15,* 156–167.

Bell, M., Bretzlauf, J., Dillard, A., Hartfield, R., Isaccs, A., McBride, J., . . . Saecker, P. (2007). *Everyday mathematics: Teacher's lesson guide, grade 6 volume 1* (3rd ed.). Chicago, IL: McGraw Hill.

National Governors Association Center for Best Practices, & Council of Chief State School Officers. (2010). *Common core state standards for mathematics.* Washington, DC: Authors.

Sheffield, L. J. (2003). *Extending the challenge in mathematics: Developing mathematical promise in K–8 students.* Thousand Oaks, CA: Corwin Press.

About the Author

Jerry Burkhart has been teaching and learning math with gifted students in Minnesota for nearly 20 years. He has degrees in physics, mathematics, and math education from University of Colorado, Boulder, and Minnesota State University, Mankato. Jerry provides professional development for teachers and is a regular presenter at conferences addressing topics of meeting the needs of gifted students in mathematics.

Common Core State Standards Alignment

Exploration	Common Core State Standards in Mathematics
Exploration 1: Building Blocks	4.OA.B Gain familiarity with factors and multiples.
Exploration 2: 1,000 Lockers	4.OA.B Gain familiarity with factors and multiples.
Exploration 3: Factoring Large Numbers	4.OA.B Gain familiarity with factors and multiples.
Exploration 4: Factor Scramble	4.OA.B Gain familiarity with factors and multiples.
Exploration 5: How Many Factors?	4.OA.B Gain familiarity with factors and multiples. 6.EE.A Apply and extend previous understandings of arithmetic to algebraic expressions. 8.EE.A Work with radicals and integer exponents.
Exploration 6: Differences and Greatest Common Factors	6.NS.B Compute fluently with multi-digit numbers and find common factors and multiples. 4.NF.A Extend understanding of fraction equivalence and ordering.
Exploration 7: A Measurement Dilemma	6.NS.B Compute fluently with multi-digit numbers and find common factors and multiples. 6.EE.A Apply and extend previous understandings of arithmetic to algebraic expressions.
Exploration 8: Paper Pool	6.NS.B Compute fluently with multi-digit numbers and find common factors and multiples. 6.RP.A Understand ratio concepts and use ratio reasoning to solve problems.
Exploration 9: The GCF-LCM Connection	6.NS.B Compute fluently with multi-digit numbers and find common factors and multiples.
Exploration 10: Mathematical Mystery Code	4.OA.B Gain familiarity with factors and multiples. 6.EE.A Apply and extend previous understandings of arithmetic to algebraic expressions. 7.EE.A Use properties of operations to generate equivalent expressions. 8.EE.A Work with radicals and integer exponents.

Note: Please see p. 8 of the book for details on how to connect and extend the core learning of content in these lessons.